Lecture Notes in Computer Science   **10080**

Commenced Publication in 1973
Founding and Former Series Editors:
Gerhard Goos, Juris Hartmanis, and Jan van Leeuwen

More information about this series at http://www.springer.com/series/7408

Raghunath Nambiar · Meikel Poess (Eds.)

Performance Evaluation and Benchmarking

Traditional - Big Data - Internet of Things

8th TPC Technology Conference, TPCTC 2016
New Delhi, India, September 5–9, 2016
Revised Selected Papers

 Springer

Editors
Raghunath Nambiar
Cisco Systems, Inc.
San Jose, CA
USA

Meikel Poess
Oracle Corporation
Redwood City, CA
USA

ISSN 0302-9743   ISSN 1611-3349   (electronic)
Lecture Notes in Computer Science
ISBN 978-3-319-54333-8   ISBN 978-3-319-54334-5   (eBook)
DOI 10.1007/978-3-319-54334-5

Library of Congress Control Number: 2017932124

LNCS Sublibrary: SL2 – Programming and Software Engineering

Printed on acid-free paper

This Springer imprint is published by Springer Nature
The registered company is Springer International Publishing AG
The registered company address is: Gewerbestrasse 11, 6330 Cham, Switzerland

The original version of the cover and title page were revised. The title was corrected from "Performance Evaluation and Benchmarking. Traditional - Big Data - Interest of Things" to "Performance Evaluation and Benchmarking. Traditional - Big Data - Internet of Things". An erratum to cover and frontmatter can be found at DOI: 10.1007/978-3-319-54334-5_11

Preface

The Transaction Processing Performance Council (TPC) is a non-profit organization established in August 1988. Over the years, the TPC has had a significant impact on the computing industry's use of industry-standard benchmarks. Vendors use TPC benchmarks to illustrate performance competitiveness for their existing products, and to improve and monitor the performance of their products under development. Many buyers use TPC benchmark results as points of comparison when purchasing new computing systems.

The information technology landscape is evolving at a rapid pace, challenging industry experts and researchers to develop innovative techniques for evaluation, measurement, and characterization of complex systems. The TPC remains committed to developing new benchmark standards to keep pace with these rapid changes in technology. One vehicle for achieving this objective is the TPC's sponsorship of the Technology Conference Series on Performance Evaluation and Benchmarking (TPCTC) established in 2009. With this conference series, the TPC encourages researchers and industry experts to present and debate novel ideas and methodologies in performance evaluation, measurement, and characterization.

This book contains the proceedings of the 8th TPC Technology Conference on Performance Evaluation and Benchmarking (TPCTC 2016), held in conjunction with the 41st International Conference on Very Large Data Bases (VLDB 2016) in New Delhi, India, during September 5–9, 2016, including selected peer-reviewed papers as well as an invited paper and a keynote paper.

The hard work and close cooperation of a number of people contributed to the success of this conference. We would like to thank the members of TPC and the organizers of VLDB 2016 for their sponsorship; the members of the Program Committee and Publicity Committee for their support; and the authors and the participants, who are the primary reason for the success of this conference.

January 2017

Raghunath Nambiar
Meikel Poess

TPCTC 2016 Organization

General Chairs

Raghunath Nambiar	Cisco, USA
Meikel Poess	Oracle, USA

Program Committee

Alain Crolotte	Teradata, USA
Akon Dey	University of Sydney, Australia
Berni Schiefer	IBM, Canada
Chaitanya Baru	SDSC, USA
Daniel Bowers	Gartner, USA
Dhabaleswar Panda	The Ohio State University, USA
Francois Raab	Infosizing, USA
Harumi Kuno	HP Labs, USA
Marco Vieira	University of Coimbra, Portugal
Michael Brey	Oracle, USA
Paul Cao	HP, USA
Reza Taheri	VMWare, USA
Tilmann Rabl	University of Toronto, Canada
Yanpei Chen	Splunk, USA

Publicity Committee

Raghunath Nambiar	Cisco, USA
Andrew Bond	Red Hat, USA
Miso Cilimdzic	Microsoft, USA
Meikel Poess	Oracle, USA
Reza Taheri	VMware, USA
Michael Majdalany	L&M Management Group, USA
Forrest Carman	Owen Media, USA
Andreas Hotea	Hotea Solutions, USA

About the TPC

Introduction to the TPC

The Transaction Processing Performance Council (TPC) is a non-profit organization that defines transaction processing and database benchmarks and distributes vendor-neutral performance data to the industry. Additional information is available at http://www.tpc.org/.

TPC Memberships

Full Members

Full Members of the TPC participate in all aspects of the TPC's work, including development of benchmark standards and setting strategic direction. The Full Member application can be found at http://www.tpc.org/information/about/app-member.asp.

Associate Members

Certain organizations may join the TPC as Associate Members. Associate Members may attend TPC meetings, but are not eligible to vote or hold office. Associate membership is available to non-profit organizations, educational institutions, market researchers, publishers, consultants, governments, and businesses that do not create, market, or sell computer products or services. The Associate Member application can be found at http://www.tpc.org/information/about/app-assoc.asp.

Academic and Government Institutions

Academic and government institutions are invited join the TPC and a special invitation can be found at http://www.tpc.org/information/specialinvitation.asp.

Contact the TPC

TPC
Presidio of San Francisco
Building 572B (surface)
P.O. Box 29920 (mail)
San Francisco, CA 94129-0920
USA
Voice: 415-561-6272
Fax: 415-561-6120
E-mail: info@tpc.org

How to Order TPC Materials

All of our materials are now posted free of charge on our website. If you have any questions, please feel free to contact our office directly or by e-mail at info@tpc.org.

Benchmark Status Report

The TPC Benchmark Status Report is a digest of the activities of the TPC and its technical subcommittees. Sign-up information can be found at the following URL: http://www.tpc.org/information/about/email.asp.

TPC 2016 Organization

Full Members

Actian
Cisco
Cloudera
Dell
DataCore
Fujitsu
HP Enterprise
Hitachi
Huawei
IBM
Inspur
Intel
Lenovo
Microsoft
Oracle
Pivotal
Red Hat
SAP
Teradata
VMware

Associate Members

IDEAS International
San Diego Super Computing Center
Telecommunications Technology Association
University of Coimbra, Portugal
CAICT

Steering Committee

Andrew Bond (Red Hat)
Michael Brey (Oracle)
Matthew Emmerton (HP)
Raghunath Nambiar (Cisco)
Jamie Reding (Microsoft)

Public Relations Committee

Andrew Bond (Red Hat)
Raghunath Nambiar (Cisco), Chair
Miso Cilimdzic, Microsoft, USA
Meikel Poess (Oracle)
Reza Taheri (VMware)

Technical Advisory Board

Andrew Bond (Red Hat)
Paul Cao (HP)
Matthew Emmerton (IBM)
John Fowler (Oracle)
Jamie Reding (Microsoft), Chair
Nicholas Wakou (Dell)

Technical Subcommittees and Chairs

TPC-ACID-AR: John Fowler
TPC-Pricing: Jamie Reding
TPC-C: Jamie Reding
TPC-DI: Meikel Poess
TPC-DS: Meikel Poess
TPC-E: Matthew Emmerton
TPC-H: Miso Cilimdzic
TPCx-V: Reza Taheri
TPC-VMS: Reza Taheri
TPCx-BB: Bhaskar Gowda
TPCx-HS: Tariq Magdon-Ismail

Working Groups and Chairs

TPC-IoT: Raghunath Nambiar

Contents

Industry Standards for the Analytics Era: TPC Roadmap

Raghunath Nambiar[1(✉)] and Meikel Poess[2]

[1] Cisco Systems, Inc., 275 East Tasman Drive, San Jose, CA 95134, USA
rnambiar@cisco.com
[2] Oracle Corporation, 500 Oracle Parkway, Redwood Shores, CA 94065, USA
meikel.poess@oracle.com

Abstract. The Transaction Processing Performance Council (TPC) is a non-profit organization focused on developing data-centric benchmark standards and disseminating objective, verifiable performance data to industry. This paper provides a high-level summary of TPC benchmark standards, technology conference initiative, and new development activities in progress.

Keywords: Industry standards · Database benchmarks

1 TPC Benchmark Timelines

Founded in 1988, the Transaction Processing Performance Council (TPC) is a non-profit corporation dedicated to creating and maintaining benchmarks which measure database performance in a standardized, objective and verifiable manner. Looking back to the 1980s, many companies practiced something known as "benchmarketing" – a practice in which organizations made performance claims based on internal benchmarks. The goal of running tailored benchmarks was simply to make one specific company's solution look far superior to that of the competition, with the objective of increasing sales. Companies created configurations specifically designed to maximize performance, called "benchmark specials," to force comparisons between non-comparable systems.

In response to this growing practice, a small group of individuals became determined to find a fair and neutral means to compare performance across database systems. Both influential academic database experts and well-known industry leaders contributed to this effort. Their important work on the topic eventually led to the creation of the TPC. Today 18 full members and five associate members comprise the TPC.

The most critical contribution of the TPC has been providing the industry with methodologies for calculating overall system-level performance and price for performance [1, 2].

Over the years the TPC has changed its mission – from defining transaction-processing benchmarks (when founded in 1988), to defining transaction processing benchmarks and database benchmarks (1999), and now defining data centric benchmarks inline with industry trends (2015) [1, 2].

To date the TPC has approved a total of sixteen independent benchmarks. Of these benchmarks, nine are currently active: TPC-C, TPC-H, TPC-E, TPC-DS, TPC-VMS, TPC-DI and TPCx-HS. New benchmarks are under development is TPC-IoT. See Fig. 1 for the benchmark timelines.

© Springer International Publishing AG 2017
R. Nambiar and M. Poess (Eds.): TPCTC 2016, LNCS 10080, pp. 1–6, 2017.
DOI: 10.1007/978-3-319-54334-5_1

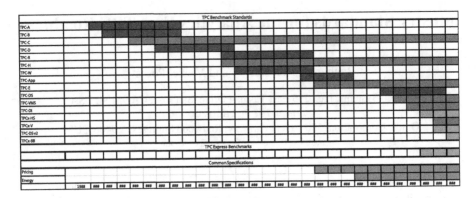

Fig. 1. TPC benchmark timelines

A high-level summary of current active standard are listed below:

- Transaction Processing
  - TPC-C: TPC-C simulates a complete computing environment where a population of users executes transactions against a database. While the benchmark portrays the activity of a wholesale supplier, TPC-C is not limited to the activity of any particular business segment, but rather represents any industry that must manage, sell or distribute a product or service.
  - TPC-E: The TPC-E benchmark uses a database to model a brokerage firm with customers who generate transactions related to trades, account inquiries and market research. The brokerage firm in turn interacts with financial markets to execute orders on behalf of the customers and updates relevant account information.
- Decision Support
  - TPC-H: An ad-hoc, decision support benchmark widely popular in industry and academia. Vendors continue to publish results on single node configurations as well as large scale-out configurations.
  - TPC-DS: A complex decision support benchmark representative of modern decision support systems. TPC took several years to develop this benchmark and reach consensus approval as a standard. No official publications have been made. TPC-DS 2.0 is under development. A major change is removing the relational database properties to support emerging platforms like Hadoop [3, 4].
  - TPC-DI: A data integration benchmark (also known as ETL) combines and transforms data extracted from a brokerage firm's OLTP system along with other sources of data, and loads it into a data warehouse. No official publications have been made [5].
- Big Data and Analytics
  - TPCx-HS: The industry's first Big Data benchmark standard is also TPC's first benchmark in the TPC Express benchmark category. The model is based on a simple application that is highly relevant to hardware and software dealing with Big Data systems in general [6].
  - TPCx-BB: TPCx-BB measures the performance of both hardware and software components by executing 30 frequently performed analytical queries in the

context of retailers with physical and online store presences. The queries are expressed in SQL for structured data and in machine learning algorithms for semi-structured and unstructured data. The SQL queries can use Hive or Spark, while the machine learning algorithms use machine learning libraries, user defined functions, and procedural programs [7].

- Virtualization
  - TPC-VMS: A single system virtualization benchmark leveraging TPC-C, TPC-E, TPC-H and TPC-DS benchmarks by adding the methodology and requirements for running and reporting performance metrics for virtualized databases [8].
  - TPCx-V: The TPCx-V benchmark measures the performance of a server running virtualized databases. It is similar to previous virtualization benchmarks in that it has many virtual machines (VMs) running different workloads. It is also similar to previous TPC benchmarks in that it uses the schema and transactions of the TPC-E benchmark. But TPCx-V is unique because, unlike previous virtualization benchmarks, it has a database-centric workload, and models many properties of cloud servers, such as multiple virtual machines running at different load demand levels, and large fluctuations in the load level of each virtual machine [8].

2 TPCTC Conference Series

To keep pace with rapid changes in technology, in 2009, the TPC initiated a conference series on performance analysis and benchmarking. The TPCTC has been challenging industry experts and researchers to develop innovative techniques for performance evaluation, measurement, and characterization of hardware and software systems. Over the years it has emerged as a leading forum to present and debate the latest and greatest in the world of benchmarking. The topics of interest included:

- Big data
- Data analytics
- Internet of Things (IoT)
- In-memory databases
- Social media infrastructure
- Security
- Hybrid workloads
- Complex event processing
- Database optimizations
- Disaster tolerance and recovery
- Energy and space efficiency
- Hardware innovations
- Cloud computing
- Virtualization
- Lessons learned in practice using TPC workloads
- Enhancements to TPC workloads
- Data integration

A short summary of the TPCTC conferences are listed below.

The first TPC Technology Conference on Performance Evaluation and Benchmarking (TPCTC 2009), held in conjunction with the 35th International Conference on Very Large Data Bases (VLDB 2009) in Lyon, France from August 24th to August 28th, 2009 [9].

The second TPC Technology Conference on Performance Evaluation and Benchmarking (TPCTC 2010) was held in conjunction with the 36th International Conference on Very Large Data Bases (VLDB 2010) in Singapore from September 13th to September 17th, 2010 [10].

The third TPC Technology Conference on Performance Evaluation and Benchmarking (TPCTC 2011), held in conjunction with the 37th International Conference on Very Large Data Bases (VLDB 2011) in Seattle, Washington from August 29th to September 3rd, 2011 [11].

The fourth TPC Technology Conference on Performance Evaluation and Benchmarking (TPCTC 2012), held in conjunction with the 38th International Conference on Very Large Data Bases (VLDB 2012) in Istanbul, Turkey from August 27th to August 31st, 2012 [12].

The fifth TPC Technology Conference on Performance Evaluation and Benchmarking (TPCTC 2013), held in conjunction with the 39th International Conference on Very Large Data Bases (VLDB 2013) in Riva del Garda, Trento, Italy from August 26th to August 30st, 2013 [13].

The sixth TPC Technology Conference on Performance Evaluation and Benchmarking (TPCTC 2014), held in conjunction with the 40th International Conference on Very Large Data Bases (VLDB 2014) in Hangzhou, China, from September 1st to September 5th, 2014 [14].

The seventh TPC Technology Conference on Performance Evaluation and Benchmarking (TPCTC 2015), held in conjunction with the 41st International Conference on Very Large Data Bases (VLDB 2015) in Kohala Coast, USA, from August 31st to September 4th, 2015 [15].

The eighth TPC Technology Conference on Performance Evaluation and Benchmarking (TPCTC 2016), held in conjunction with the 42nd International Conference on Very Large Data Bases (VLDB 2016) in New Delhi, India, from September 5th to September 9th, 2016.

TPCTC has had a significant positive impact on the TPC. TPC is able to attract new members from industry and academia to join the TPC. The formation of working groups on Big Data, Virtualization and Internet of Things (IoT) was a direct result of TPCTC conferences.

3   Outlook

TPC remains committed to develop relevant standards in collaboration with industry and research communities and continue to enable fair comparison of technologies and products in terms of performance, cost of ownership. New additions to TPC standards in recent years have been standards for Big Data and Analytics and Virtualization [7–9].

Foreseeing the industry transition to digital transformation the TPC has created a working group to develop set of standards for hardware and software pertaining to Internet of Things (IoT). Companies, research and government institutions who are interested in influencing the development of such benchmarks are encouraged to join the TPC [2].

The TPC Pricing Subcommittee has been chartered to recommend revisions to the existing pricing methodology to support the benchmark in public cloud environments.

Acknowledgements. Developing benchmark standards require a huge effort to conceptualize, research, specify, review, prototype, and verify the benchmark. The authors acknowledge the work and contributions of past and present members of the TPC.

References

1. Nambiar, R., Poess, M.: Reinventing the TPC: from traditional to big data to internet of things. In: Nambiar, R., Poess, M. (eds.) TPCTC 2015. LNCS, vol. 9508, pp. 1–7. Springer, Heidelberg (2016). doi:10.1007/978-3-319-31409-9_1
2. Nambiar, R., Poess, M.: Keeping the TPC relevant! PVLDB **6**(11), 1186–1187 (2013)
3. Nambiar, R., Wakou, N., Masland, A., Thawley, P., Lanken, M., Carman, F., Majdalany, M.: Shaping the landscape of industry standard benchmarks: contributions of the transaction processing performance council (TPC). In: Nambiar, R., Poess, M. (eds.) TPCTC 2011. LNCS, vol. 7144, pp. 1–9. Springer, Heidelberg (2012). doi:10.1007/978-3-642-32627-1_1
4. Nambiar, R., Poess, M.: The making of TPC-DS. In: VLDB 2006, pp. 1049–1058 (2006)
5. Pöss, M., Nambiar, R., Walrath, D.: Why you should run TPC-DS: a workload analysis. In: VLDB 2007, pp. 1138–1149 (2007)
6. Poess, M., Rabl, T., Caufield, B.: TPC-DI: the first industry benchmark for data integration. PVLDB **7**(13), 1367–1378 (2014)
7. Nambiar, R., Poess, M., Dey, A., Cao, P., Magdon-Ismail, T., Da Qi Ren, Bond, A.: Introducing TPCx-HS: the first industry standard for benchmarking big data systems. In: Nambiar, R., Poess, M. (eds.) Technology Conference on Performance Evaluation and Benchmarking, TPCTC 2014, pp. 1–12. Springer, Heidelberg (2014)
8. Baru, C.: Discussion of BigBench: a proposed industry standard performance benchmark for big data. In: Nambiar, R., Poess, M. (eds.) TPCTC 2014. LNCS, vol. 8904, pp. 44–63. Springer, Heidelberg (2014). doi:10.1007/978-3-319-15350-6_4
9. Bond, A., Johnson, D., Kopczynski, G., Taheri, H.R.: Profiling the performance of virtualized databases with the TPCx-V benchmark. In: Nambiar, R., Poess, M. (eds.) TPCTC 2015. LNCS, vol. 9508, pp. 156–172. Springer, Heidelberg (2016). doi:10.1007/978-3-319-31409-9_10
10. Nambiar, R., Poess, M. (eds.): Performance Evaluation and Benchmarking, TPCTC 2009. LNCS, vol. 5895. Springer, Heidelberg (2009). doi:10.1007/978-3-642-10424-4. ISBN 978-3-642-10423-7
11. Nambiar, R., Poess, M. (eds.): Performance Evaluation, Measurement and Characterization of Complex Systems, TPCTC 2010. LNCS, vol. 6417. Springer, Heidelberg (2011). doi: 10.1007/978-3-642-18206-8. ISBN 978-3-642-18205-1
12. Nambiar, R., Poess, M. (eds.): Topics in Performance Evaluation, Measurement and Characterization, TPCTC 2011. LNCS, vol. 7144. Springer, Heidelberg (2012). doi: 10.1007/978-3-642-32627-1. ISBN 978-3-642-32626-4
13. Nambiar, R., Poess, M. (eds.): Selected Topics in Performance Evaluation and Benchmarking, TPCTC 2012. LNCS, vol. 7755. Springer, Heidelberg (2013). doi:10.1007/978-3-642-36727-4. ISBN 978-3-642-36726-7

14. Nambiar, R., Poess, M. (eds.): Performance Characterization and Benchmarking, TPCTC 2013. LNCS, vol. 8391. Springer, Heidelberg (2014). doi:10.1007/978-3-319-04936-6. ISBN 978-3-319-04935-9

15. Nambiar, R., Poess, M. (eds.): Performance Characterization and Benchmarking. Traditional to Big Data, TPCTC 2014. LNCS, vol. 8904. Springer, Heidelberg (2014). doi:10.1007/978-3-319-15350-6. ISBN 978-3-319-15349-0

16. Nambiar, R., Poess, M. (eds.): Performance Evaluation and Benchmarking: Traditional to Big Data to Internet of Things, TPCTC 2015. LNCS, vol. 9508. Springer, Heidelberg (2016). doi:10.1007/978-3-319-31409-9. ISBN 978-3-319-31408-2

TPCx-HS on the Cloud!

Nicholas Wakou[1(✉)], Michael Woodside[1], Arkady Kanevsky[1],
Fazal E Rehman Khan[2], and Mofassir ul Islam Arif[2]

[1] Dell Inc., Round Rock, TX 78682, USA
{Nicholas.Wakou, C.Michael.Woodside,
Arkady.Kanevsky}@dell.com
[2] xFlow Research Inc., Software Technology Park, Islamabad, Pakistan
{Fazal.Rehman,Mofassir.Arif}@xFlowresearch.com
http://www.Dell.com
http://www.xFlowResearch.com

Abstract. The introduction of web scale operations needed for social media coupled with ease of access to the internet by mobile devices has exponentially increased the amount of data being generated every day. By conservative estimates the world generates close to 50,000 GB of data every second, 90% of which is unstructured, and this growth is accelerating. From its origins as a web log processing system at Yahoo, the open source nature and efficient processing of Apache Hadoop has made it the industry standard for Big Data processing.

TPCx-HS was the first benchmark standard by a major Industry-Standard performance consortium for the Big Data space. TPCx-HS is a derivative of Apache Hadoop Workloads; Teragen, Terasort and Teravalidate. Ever since its release by the TPC in August 2014, all the 18 results published (as of August 2016) have been based on on-premise, Bare-metal hardware configurations.

This paper will show how Hadoop can be deployed on an OpenStack cloud using the OpenStack Sahara project and how TPCx-HS can be used to measure and evaluate the performance of the Cloud under Test (CuT). It will also show how an OpenStack cloud can be optimized to get the performance of TPCx-HS on the Cloud to match as closely as possible that on a Bare-metal configuration. Lastly, it will share results and experiences based on a Hadoop on Cloud Proof-of-Concept (POC), a study that was undertaken by the Dell Open Source Solutions team.

Keywords: Apache Hadoop · OpenStack · Big data · Cloud · TPCx-HS · Benchmark

1 Introduction

The complex nature of big data is primarily due to the unstructured nature of much of the data that is generated by modern technologies such as that from web logs, RFID, sensors, and smart phones [1]. This coupled with web scale operations of companies like Google, Yahoo and Facebook exponentially increased the amount of data being generated. This was the turning point in the big data life cycle which demanded the need for a system to efficiently manage and process these large amounts of data. Hadoop emerged from the efforts of these data giants and due to the open source nature

© Springer International Publishing AG 2017
R. Nambiar and M. Poess (Eds.): TPCTC 2016, LNCS 10080, pp. 7–23, 2017.
DOI: 10.1007/978-3-319-54334-5_2

of several of its key components, quickly became the standard for managing large volumes of unstructured data. The two main components of Apache Hadoop platform are Hadoop Distributed File System (HDFS) and MapReduce. A number of major Computer companies now offer Hadoop-based solutions for analyzing big data use-cases. An industry-standard benchmark was therefore required to compare and differentiate these offerings. TPC Express Benchmark™HS (TPCx-HS) was developed to provide an objective measure of hardware, operating system and commercial Apache Hadoop File System API compatible software distributions, and to provide the industry with verifiable performance, price-performance and availability metrics [2]. Traditionally Hadoop is deployed on customer premise and on physical servers. Due to its inherently efficient nature in terms of resource allocation and utilization, there appears little need for a cloud Hadoop deployment. This trend is changing with the advancements of cloud technology. A cloud deployment offers the ability to conveniently scale the cluster as needed. Multi-tenancy is also a big advantage when it comes to facilitating multiple users on the same physical hardware. Furthermore, a cloud deployment, coupled with multi-tenancy, is greatly complimented by the resources and security segregation. Each tenant has full control over their resources without incurring any risk to the resources managed by the other tenants [3]. Hadoop can now be run in a cloud in a way that is efficient and performance can be made comparable with physical hardware after a few configurations and tweaks. This paper provides recommendations for cloud configuration and Hardware options for running Hadoop workloads in the cloud.

2   Related Work

Google searches on Hadoop on Cloud show that some work has been done on running Hadoop on a public cloud [4] and on private clouds [5]. For the most part, moving from the traditional Bare-metal deployment of Hadoop to the Cloud is still seen as a challenge by Enterprises mainly due to performance concerns in the Cloud. One of the most related studies to this paper was work conducted by Accenture on a price-performance comparison of a Bare-metal Hadoop cluster and cloud-based Hadoop clusters. Accenture used their own TCO model and the Accenture Data Platform Benchmark which provided three real-world Hadoop applications to compare the execution-time performance of the clusters [6]. The above mentioned references make the case for running Hadoop on a cloud; the advantages and challenges. They show that despite performance challenges, it still makes sense to run Hadoop on the Cloud. This paper shows that with appropriate cloud configurations and settings, Hadoop performance on the cloud can match that on Bare-metal. This study deployed TPCx-HS Big Data workloads on an OpenStack Cloud.

3   System Under Test

There were 2 main Systems under Test; Cloud and Bare-metal. The Bare-metal system consisted of $4\times$ Dell R730xd servers, the details can be found in Table 1.

Table 1. Bare-metal System under Test

Role	Model	Qty.	CPUs	Memory	Storage	Network adaptor
Name Node	Dell R730xd	1	2× Intel 12-Core E5-2690 v3	128 GB, 8 × 16 GB DIMMS, 2133 MT/s	16 × 1 TB (2.5″, 7.2 K, HDD, SAS, JBOD), 2 × 300 GB (2.5″,HDD, SAS, RAID 1)	Intel 2P 10G X520, Intel 2P 10G X520 + 2P 1G I350 rNDS
Data Node	Dell R730xd	3	2× Intel 12-Core E5-2690 v3	128 GB, 8 × 16 GB DIMMS, 2133 MT/s	16 × 1 TB (2.5″, 7.2 K, HDD, SAS, JBOD), 2 × 300 GB (2.5″, HDD, SAS, RAID 1)	Intel 2P 10G X520, Intel 2P 10G X520 + 2P 1G I350 rNDS

The Cloud system was based on the Dell Red Hat OpenStack Cloud Solution Reference Architecture (RA) version 4.0.1 based on Red Hat OpenStack Cloud Platform 7 (OSP7). Details are provided in Fig. 1. The Cloud setup used 2 types of storage configurations both shown in Fig. 1.

1. Ceph Storage for all Compute nodes; for both block and Nova ephemeral storage.
2. Local Storage on each Compute node; for both block and Nova ephemeral storage.

Sahara is an OpenStack project for deployment and management of Hadoop clusters. Sahara was used to deploy Cloudera CDH 5.3 that in turn used OpenStack APIs to manage Instances to run Hadoop. The Hadoop cluster consisted of Cloudera Manager Instance (EdgeNode), Namenode Instance (Master) and Datanode Instances (Workers). The physical resources allocated to these instances are listed in Table 2. Each Compute node was configured as a separate OpenStack availability zone in order to keep the Cloudera Manager and Namenode Instances on one Compute node and then evenly distribute the Worker Instance(s) across all Compute nodes. Unless otherwise stated, Datanode instances will be referred to as Workers.

Table 2. Resource allocation

Role	Number of instances	vCPU	Memory	Compute node
Cloudera manager instance	1	2	8 GB	Compute node 1
Namenode instance	1	6	16 GB	Compute node 1
Worker instances	3–60	40	80 GB	Compute nodes 1–3

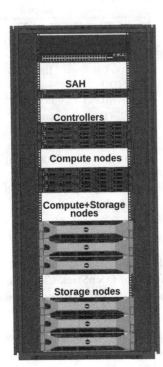

Networking
2xDell S4048   10 GbE
1xDell S3048 1GbE

SAH
1xR630
Processors
2xIntel 12-Core E5-2650v4
Memory
128GB,
16GBx8 DDR-4, 2400MT/s
Disks
4x372GB,2.5",
SAS,SSD,
RAID 10 400GB
+ RAID10 344GB, 12GBps
Network
Intel 2P 10G X520
Intel 2P 10G X520
+2P 1G I350 rNDS

Ceph
Storage nodes
3xR730xd
Each node:
Processors
2xIntel 12-Core E5-2650v4
Memory
128 GB, 8x 16GB, DDR-4, 2400 MT/s
Disks
13x2TB, 3.5",7.2K,HDD,
SAS, JBOD
2x300GB,2.5",15K,HDD,SAS,
RAID1 278GB
3x200GB,2.5" SSD
Network
Intel 2P 10G X520
Intel 2P 10G X520
+ 2P 1G I350 rNDS

Controllers
3xR630
Processors
2xIntel 12-Core E5-2650v4
Memory
128GB,16GBx8 DDR-4 ,2400MT/s
Disks
4x600GB,2.5",SAS,HDD,10K,
RAID 10 1.2TB
Network
Intel 2P 10G X520
Intel 2P 10G X520
+ 2P 1G I350 rNDS

Compute nodes
3xR630
Each node:
Processors
2xIntel 12-Core E5-2650v4
Memory
128 GB, 8x 16GB, DDR-4, 2400 MT/s
Disks
8x600 GB,2.5",10K, HDD SAS
RAID10 2.2TB
Network
Intel 2P 10G X520
Intel 2P 10G X520
+ 2P 1G I350 rNDS

Compute + Local
Storage nodes
3xR730xd
Each node:
Processors
2x Intel 12-Core E5-2690 v3
Memory
128 GB, 8x 16GB, DDR-4, 2133 MT/s
Disks
16x1TB, 2.5". 7.2K, HDD,
SAS JBOD
2x300 GB, 2.5", HDD, SAS
RAID 1
Network
Intel 2P 10G X520
Intel 2P 10G X520
+ 2P 1G I350 rNDS

Fig. 1. Dell Red Hat OpenStack Platform

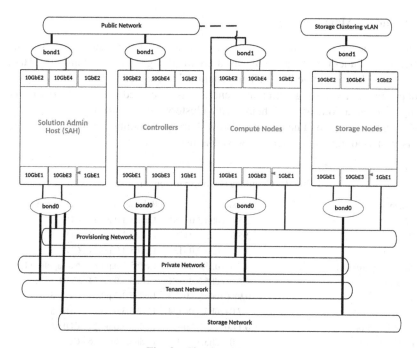

Fig. 2. Cloud architecture

3.1 OpenStack Sahara

OpenStack Sahara provides a robust interface to easily provision and scale Hadoop clusters. As an OpenStack component, OpenStack Sahara is fully integrated into the OpenStack ecosystem; for example, users can administer the entire Hadoop data processing workflow through the OpenStack dashboard (Horizon) – from configuring clusters, all the way to launching and running jobs on them [7].

A cluster deployed by Sahara consists of node groups. Node groups vary by their role, parameters and number of machines. In order to simplify cluster provisioning, Sahara makes use of two kinds of templates: node group templates and cluster templates. A cluster template is made up of multiple node group templates, while a node group template specifies the role and services that are needed in that group. The use of these templates reduces the cluster deployment and configuration time.

Sahara uses different plugins to provision a specific data processing frame-work or Hadoop distribution. There are several supported plugins for the OpenStack Kilo release including Vanilla, Cloudera, Ambari and Spark. We used the Cloudera (CDH) 5.3.0 plugin which allows the deployment and operation of a cluster with Cloudera Manager.

OpenStack deploys instances of the cluster based on a pre-built image with an installed OS. The image requirements for Sahara depend on the plugin and data processing framework version. With the Cloudera 5.3.0 plugin, we used the Centos 6.6 based image with preinstalled packages of CDH 5.3.0 [8].

3.2   Cloudera Manager 5.3

Cloudera Manager makes it easy to manage Hadoop deployments of any scale in production. Cloudera Manager 5.3 was used to configure and monitor the Hadoop clusters through its intuitive UI [9]. This functionality enabled us to easily configure the Hadoop clusters for testing different scenarios as discussed in Sect. 5: Performance Testing. Cloudera Manager UI helps with clusterwide monitoring of all hosts and services, cluster usage and health, and also with troubleshooting problems. The version summary of Hadoop components deployed is listed in Table 3.

Table 3. Cloudera Hadoop version summary

Component	Version	Release	CDH version
YARN	2.5.0+cdh5.3.0+781	1.cdh5.3.0.p0.54	CDH 5
HDFS	2.5.0+cdh5.3.0+781	1.cdh5.3.0.p0.54	CDH 5
hue-common	3.7.0+cdh5.3.0+134	1.cdh5.3.0.p0.24	CDH 5
Keytrustee Keyprovider	5.5.0+cdh5.5.0+0	1.cdh5.5.0.p0.1	Not applicable
hadoop-kms	2.5.0+cdh5.3.0+781	1.cdh5.3.0.p0.54	CDH 5
HBase	0.98.6+cdh5.3.0+73	1.cdh5.3.0.p0.25	CDH 5
Hue	3.7.0+cdh5.3.0+134	1.cdh5.3.0.p0.24	CDH 5
Crunch (CDH 5 only)	0.11.0+cdh5.3.0+16	1.cdh5.3.0.p0.24	CDH 5
Llama (CDH 5 only)	1.0.0+cdh5.3.0+0	1.cdh5.3.0.p0.26	CDH 5
HttpFS	2.5.0+cdh5.3.0+781	1.cdh5.3.0.p0.54	CDH 5
Hadoop	2.5.0+cdh5.3.0+781	1.cdh5.3.0.p0.54	CDH 5
sentry	1.4.0+cdh5.3.0+126	1.cdh5.3.0.p0.26	CDH 5
MapReduce 2	2.5.0+cdh5.3.0+781	1.cdh5.3.0.p0.54	CDH 5
Lily HBase Indexer	1.5+cdh5.3.0+23	1.cdh5.3.0.p0.18	CDH 5
Flume NG	1.5.0+cdh5.3.0+79	1.cdh5.3.0.p0.18	CDH 5
Cloudera Manager Management Daemons	5.3.0	1.cm530.p0.166	Not applicable
Supervisord	3.0-cm5.3.0	Unavailable	Not applicable
Java 7	jdk1.7.0 67-cloudera	Unavailable	Not applicable
Cloudera Manager agent	5.3.0	1.cm530.p0.166	Not applicable

3.3   TPCx-HS

The results of this POC were derived from the TPCx-HS benchmark and as such are not comparable to published TPCx-HS results. TPCx-HS proved to be a viable option because of its focus on big data. TPCx-HS was developed to provide an objective measure of hardware, operating system and commercial Apache Hadoop File System API compatible software distributions, and to provide the industry with verifiable performance, price-performance and availability metrics [2]. Each run of the benchmark consisted of 2 iterations of HSGen, HSDataCheck, HSSort, HSValidate and provided us with the job run time.

4 Configurations

4.1 Hardware Configurations

Below are the hardware configurations that were used for different test scenarios discussed in Sect. 5. Table 4 shows the hardware configurations for Instance and Over-Subscription tests while Table 5 shows the hardware configurations for HDFS on local Storage, CPU Pinning/NUMA with HDFS on Local Storage and Disk and CPU Pinning/NUMA with HDFS on Local Storage tests. It should be noted that the hardware configurations used for this study are under the control of the Cloud and its administrator and are not visible to the user running the tests described in Sect. 5.

Table 4. Hardware configuration for instance and over-subscription tests

	Controller nodes	Compute nodes	Ceph storage nodes	
Server model	Dell R630	Dell R630	Dell R730xd	
CPU	Intel E5-2650 v4	Intel E5-2650 v4	Intel E5-2650 v4	
Memory	128 GB, 2400 MT/s	128 GB, 2400 MT/s	128 GB, 2400 MT/s	
BIOS version	2.0.1	2.0.1	2.0.1	
Firmware version	2.30.30.30	2.30.30.30	2.30.30.30	
HDD	4 × 600 GB, 2.5″, SAS, HDD	8 × 600 GB, 2.5″, SAS, HDD	2 × 300 GB, 2.5″, SAS, HDD	3 × 200 GB, 2.5″, SAS, SSD +13 × 2 TB, 3.5″, SAS, HDD
HDD configuration	H730 RAID 10 with drives 0, 1, 2, 3;	H730 RAID 10 with drives 0, 1, 2, 3, 4, 5, 6, 7;	H730 RAID 1 with flex bay drives 12, 13;	JBOD drives 0, 1, 2, 3, 4, 5, 6, 7, 8, 9, 10, 11, 14, 15, 16, 17
Read policy	Adaptive read ahead	Adaptive read ahead	Adaptive read ahead	–
Write policy	Write back	Write back	Write back	–
Disk cache policy	Default	Default	Default	–

Table 5. H/W configuration for HDFS on local storage, CPU & Disk pinning tests

	Controller nodes	Compute with local storage nodes	
Server model	Dell R630	Dell R730xd	
CPU	Intel E5-2650 v4	Intel E5-2690 v3	
Memory	128 GB, 2400 MT/s	128 GB, 2133 MT/s	
BIOS version	2.0.1	2.0.1	
Firmware version	2.30.30.30	2.30.30.30	
HDD	4 × 600 GB, 2.5″, SAS, HDD	2 × 300 GB, 2.5″, SAS, HDD	16 × 1 TB, 2.5″n, SAS, HDD
HDD configuration	H730 RAID 10 with drives 0, 1, 2, 3;	H730 RAID 1 with flex bay drives 24, 25;	JBOD drives 0, 1, 2, 3, 4, 5, 6, 7, 8, 9, 10, 11, 12, 13, 14, 15
Read policy	Adaptive read ahead	Adaptive read ahead	–
Write policy	Write back	Write back	–
Disk cache policy	Default	Default	–

4.2 Hadoop Configurations

In order to implement different test scenarios, several of the Hadoop configurations were varied. Some of the configurations that remained common for all the tests are listed in Table 6 while the configurations that varied from test to test to find the optimum performance are listed in Table 7.

Table 6. Common Hadoop configurations

S. No.	Configuration name	Value
1	dfs.replication	3
2	dfs.blocksize	512 MB
3	mapreduce.map.cpu.vcores	1
4	mapreduce.map.memory.mb	1024 MB
5	mapreduce.reduce.cpu.vcores	1
6	mapreduce.reduce.memory.mb	2048 MB
7	yarn.scheduler.minimum-allocation-vcores	1
8	yarn.scheduler.minimum-allocation-mb	1024 MB
9	yarn.scheduler.increment-allocation-vcores	1
10	yarn.scheduler.increment-allocation-mb	512 MB
11	yarn.app.mapreduce.am.resource.mb	2048 MB
12	mapreduce.map.sort.spill.percent	0.8
13	mapreduce.task.io.sort.mb	256 MB
14	mapreduce.job.reduce.slowstart.completedmaps	0.8

Table 7. Variable Hadoop configurations

S. No.	Configuration name	Range
1	yarn.nodemanager.resource.memory-mb	4 GB–96 GB
2	yarn.nodemanager.resource.cpu-vcores	2–40
3	yarn.scheduler.maximum-allocation-mb	4 GB–96 GB
4	yarn.scheduler.maximum-allocation-vcores	2–40

5 Performance Testing

Based on prior research and virtualization experience [10], a few OpenStack configurations that were known to have a high impact on performance were selected for performance testing. These included over-subscription, use of local storage on Compute nodes, NUMA nodes and Disk Pinning. Test cases were developed to run and measure the performance of TPCx-HS workloads on these configurations.

5.1 Instance Configuration Tests

The goal of the instance tests was to understand the impact of Instance (VM) configurations on TPCx-HS performance.

First, the number of workers versus the resources allocated to them was tested. In essence, it was necessary to determine whether a small number of instances, each with a higher resource allocation, would work better than a greater number of worker instances with relatively fewer resources allocated to each of them. In order to test the above scenarios, an arrangement (I1...I5) of vCPU, memory and number of instances was specified as shown in Table 8. TPCx-HS tests were executed on Hadoop clusters based on each of the specified instance configurations.

Table 8. Instance configuration tests

Test ID	Nova flavor	Worker per Node/Total	vCPU per Worker	Memory per Worker	Storage per Worker	Storage type
I1	custom	1/3	40	96 GB	16 TB	Ceph
I2	m1. medium	20/60	2	4 GB	1 TB	Ceph
I3	m1. large	10/30	4	8 GB	2 TB	Ceph
I4	m1. xlarge	5/15	8	16 GB	4 TB	Ceph
I5	m1. cxlarge	4/12	10	20 GB	4 TB	Ceph

These clusters were deployed using OpenStack Sahara. Note that m1.cxlarge was a custom flavor. These tests were run without resource over-subscription. In all tables, unless otherwise stated, Node refers to Compute Node.

While the Bare-metal hardware configuration shown in Table 1 was different from the cloud virtual machine configuration, its performance served as a datum point for comparison of results. The relative performance on the y-axis of Figs. 3 and 4 is performance compared to Bare-metal Hadoop performance datum. For the Instance configuration iterative test it was found that instance configuration I5, as shown in Table 8, provided best performance. Figure 3 shows that large-sized flavor configurations perform better. The only exception is the single-instance custom flavor which was configured with the largest size but performed poorest.

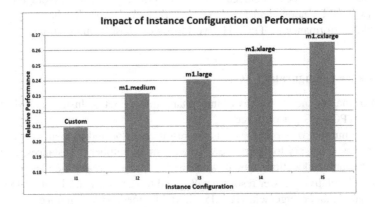

Fig. 3. Performance per instance configuration

Secondly the number of Instances per node were varied from 1 to 20 while ensuring that the number of vCPUs and YARN containers created remained constant. It was observed that performance peaks at 4 instances per node and then it tapers downwards. The result for this can be seen in Fig. 4 which shows performance with maximum number of instances for each flavor type.

The results of the instance configuration tests show that provisioning 4 m1.cxlarge configuration instances and allocating them the maximum amount of memory and vCPU resources provides the best performance. From these results, the TPCx-HS performance gain due to an optimal instance configuration can go up to 5%. The optimal instance configuration (I5) determined in this test is used in subsequent tests.

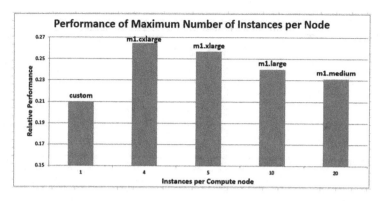

Fig. 4. Performance per number of instances

5.2   Over-Subscription Tests

In this section, the impact of resource over-subscription on aggregate performance was determined. Over-subscription was controlled by a Cloud administrator through the choice of Nova flavor.

CPU Over-Subscription Tests. In these tests, the number of instances and memory per instance was fixed, while the vCPUs assigned to each instance were increased according to the ratio shown in Table 9. For the test ID C1, the optimal configuration obtained in Sect. 5.1, I5, was used. The iterative tests for the vCPU over-subscription can be found below in Table 9.

Table 9. CPU over-subscription tests

Test ID	Ratio	Nova flavor	Worker per Node/Total	vCPU per Worker	Memory per Worker	Storage per Worker	Storage type
C1	1:1	custom	4/12	10	20 GB	4 TB	Ceph
C2	1:2	custom	4/12	20	20 GB	4 TB	Ceph
C3	1:3	custom	4/12	30	20 GB	4 TB	Ceph
C4	1:4	custom	4/12	40	20 GB	4 TB	Ceph

The number of worker instances and memory per instance values were fixed while the vCPUs per instance were increased to find optimal performance as shown in Table 9. The performance of each test was normalized by the performance of Test ID "C1". 1:1 CPU subscription yielded the best performance and Fig. 5 shows that while there is a performance cost with over-subscription, it is not so drastic. An over-subscription ratio of 1:2 results in a 2.5% drop in performance and a 1:4 results in a 23% drop.

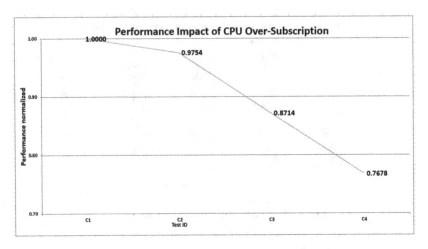

Fig. 5. CPU over-subscription

Memory Over-Subscription Tests. In these tests, the optimal arrangement of instances and vCPUs from the CPU over-subscription tests was used, while the memory per worker was increased according to the ratio shown in Table 10.

Table 10. Memory over-subscription tests

Test ID	Ratio	Nova flavor	Worker per Node/Total	vCPU per Worker	Memory per Worker	Storage per Worker	Storage type
M1	1:1	custom	4/12	10	20 GB	4 TB	Ceph
M1.1	1:1.1	custom	4/12	10	22 GB	4 TB	Ceph
M1.2	1:1.2	custom	4/12	10	24 GB	4 TB	Ceph
M1.3	1:1.3	custom	4/12	10	26 GB	4 TB	Ceph

The performance of each test was normalized by the performance of Test ID "M1" that has no over-subscription. As the ratio of vMem (virtual) to pMem (physical) was raised by 10% through 30%, Fig. 6 shows that there was a drastic drop in performance with memory over-subscription. A 10% memory over-subscription results in a 65% drop in performance while a 30% over-subscription results in a 70% drop. This test demonstrates that memory over-subscription should be avoided if performance is a consideration. Based on over-subscription tests, it was determined that memory over-subscription has a bigger impact on performance than CPU over-subscription.

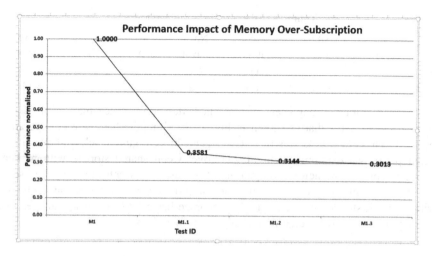

Fig. 6. Memory over-subscription

5.3 HDFS on Local Storage Tests

Some phases of the TPCx-HS workloads particularly HSGen (Teragen) are IO intensive. It was therefore important to use a storage configuration that would provide better performance. In that respect, tests were undertaken to compare HDFS performance on local storage to Ceph shared backend.

For HDFS on local storage tests, 3 Dell R730xd servers each with 16 × 1 TB SAS data drives were added to the OpenStack Cloud. See Fig. 1 and Table 5 for the detailed hardware configuration. Notice that for these Local Storage tests the same hardware was used for Bare-metal and Cloud runs, thus it is a true comparison between Hadoop on Bare-metal and Hadoop on OpenStack cloud performance. Each physical node was configured as both OpenStack Compute and Cinder Volume node. On each server, all 16 SAS drives were added to a single volume group and then OpenStack Cinder was configured to use that volume group with LVM iSCSI Volume Driver [11]. Each node was configured as a separate availability zone to ensure that volume attachment comes from the same zone and the cinder volumes are local to a Compute node.

A cluster of 12 worker instances (4 on each R730xd) was deployed and each instance was attached to 4 × 1 TB Cinder volumes for HDFS provided by the local volume server. Similarly, for Ceph-based configuration the same number (4) and size (1 TB) of volumes were attached to each Instance. HDFS replication was set at 3 for all the tests while Ceph replication was set at 1 to maintain the total replication factor of 3. See Table 11 below for test configurations.

Table 11. HDFS on local storage tests

Test ID	Ceph replicas	HDFS replication	Nova flavor	Nova flavor	Nova flavor	Nova flavor	Nova flavor
Ceph	1	3	custom	4/12	10	20 GB	4 TB
Local	–	3	custom	4/12	10	20 GB	4 TB

The local storage arrangement resulted in better performance than Ceph shared storage backend. Note that as shown in Fig. 1, the Compute hardware configuration used for Ceph storage tests was of a newer generation. Figure 7 below shows a significant performance improvement of 22% over Ceph shared storage with replication = 1 in spite of the newer Compute hardware used for Ceph Storage. The use of local storage minimizes network traffic and improves performance in an IO-bound environment. It should be noted that Ceph Storage has a lot of other advantages like resiliency that might out-weigh performance considerations.

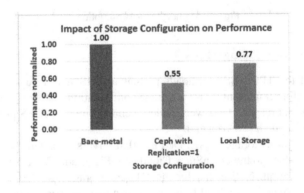

Fig. 7. HDFS on local storage

5.4 CPU Pinning/NUMA with HDFS on Local Storage Tests

The goal of this test is to understand the impact of NUMA awareness of the OpenStack scheduler on performance. To test the performance of using dedicated vCPUs with NUMA awareness, the 3 Compute Nodes (R730xds) in Sect. 5.3 above were configured to support the pinning of virtual machine instances to dedicated physical cores [12]. A cluster of 12 worker instances (4 on each R730xd) was then deployed where the vCPUs of each instance pins and exclusively use the physical CPUs. Each instance was attached to 4 × 1 TB Cinder volumes provided by the local volume server. See Table 12 below for the test configurations. Test ID "Non-NUMA" in Table 12 used the same configuration as HDFS on Local Storage described in Sect. 5.3.

Table 12. CPU pinning/NUMA with HDFS on local storage tests

Test ID	Nova Flavor	Worker per Node/Total	vCPU per Worker	Memory per Worker	Storage per Worker	Storage type
Non-NUMA	custom	4/12	10	20 GB	4 TB	Local
NUMA	custom	4/12	10	20 GB	4 TB	Local

An additional 2% performance improvement was observed by implementing CPU pinning in a configuration with HDFS on Local storage as shown in Fig. 8. The ability for the OpenStack scheduler to be aware of the underlying NUMA architecture typically optimizes the performance of individual Instances. In this test, processor affinity (CPU pinning) did not have a significant impact on performance. This could be attributed to the effects of the KVM hypervisor and should be a subject of further investigation.

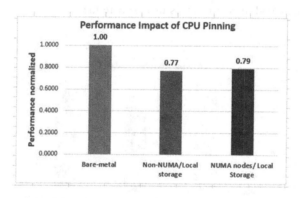

Fig. 8. CPU pinning/NUMA with HDFS on local storage

5.5   Disk and CPU Pinning/NUMA with HDFS on Local Storage Tests

The goal of this test is to understand the impact of disk pinning on performance. To test the performance of using dedicated disks as compared to shared local disks, 3 Dell R730xds were configured to implement the pinning of physical disks to the virtual machine instances. Each server had 16 × 1 TB SAS data drives. 16 separate volume groups were created and one physical disk was assigned to each volume group. OpenStack Cinder was configured to use all of those volume groups, each as an individual storage backend with LVM iSCSI Volume Driver. A cluster of 12 worker instances (4 on each R730xd) were deployed and each instance was attached to 4 × 1 TB Cinder volumes provided by one of the volume groups from the local volume server [13]. Additionally the vCPUs of each instance were pinned to the physical CPUs. See Table 13 below for the tests configurations. Test ID "NUMA" used the same configuration of "NUMA" as described in Sect. 5.4.

Table 13. Disk and CPU pinning/NUMA with HDFS on local storage tests

Test ID	Nova flavor	Worker per Node/Total	vCPU per Worker	Memory per Worker	Storage per Worker	Storage type
NUMA	custom	4/12	10	20 GB	4 TB	Local
NUMA & Disk	custom	4/12	10	20 GB	4 TB	Local

Figure 9 below shows that a performance improvement of 15% is attributed to disk pinning. TPCx-HS workloads are IO-bound during HSGen (Teragen) and shuffle phase of HSSort (Terasort). In an IO-intensive environment, the ability by instances to access and pin physical disks directly has a significant performance impact as this test has shown. Further performance gains can be achieved by use of a raw device driver instead of LVM. This test also shows that by implementing disk pinning in a configuration that uses NUMA nodes with HDFS on local storage, TPCx-HS performance on the OpenStack cloud almost matches that on bare metal.

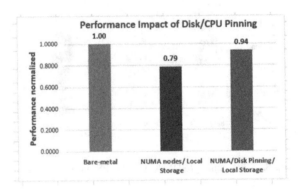

Fig. 9. Disk and CPU pinning/NUMA with HDFS on local storage

6   Conclusions

Hadoop-on-Cloud POC has shown that it is possible for the performance of Big Data workloads (like TPCx-HS on OpenStack) on the Cloud to match that on Bare-Metal. This improvement was achieved by using an optimal Instance configuration that was deployed on local storage and with the implementation of CPU and disk pinning. More performance gains can be realized by implementing the use of a raw device driver by Cinder instead of LVM used in this study. The net effect of the aggregation of optimizations shown in this paper and those that have been recommended should lead to better TPCx-HS performance on the Cloud than on Bare-metal. That has been shown to be possible in virtualized environments [10] and from the results of this POC it should be possible on the Cloud. Follow-up tests to this POC will strive to identify even more optimizations. It is worth noting that in this paper, our recommendations for OpenStack configuration and hardware choices were considered from a performance perspective. In a production environment, Openstack and Hadoop data protection best-practices should be considered. This includes use of persistent storage and raising the HDFS replication factor to greater than the default (>3).

Acknowledgments. The authors would like to thank John Terpstra, Michael Pittaro, Randy Perryman, Michael Tondee and David Grimes for participating in the technical review meetings of the POC. Their input, feedback and guidance helped shape this investigation. Mr. Ashok Malani is recognized for his technical leadership of the xFlow Research team that did such a tremendous job performing the tests and drafting this paper.

References

1. Navint: Why is big data important? (2012). www.navint.com/images/Navint.BigData. FINAL.pdf
2. TPC: Tpcx-hs (2016). http://www.tpc.org/tpcx-hs/
3. VMware: Virtualized hadoop performance with vmware vsphere 6 on highperformance servers (2015). http://www.vmware.com/files/pdf/techpaper/Virtualized-Hadoop-Performance-with-VMware-vSphere6.pdf
4. Stata, R.: Understanding hadoop-as-a-service offerings (2014). http://www.datacenterknowledge.com/archives/2014/05/14/understanding-hadoop-service-offerings/
5. Hurtgen, A.: Using apache hadoop on rackspace private cloud (2013). https://support.rackspace.com/how-to/apache-hadoop-on-rackspace-private-cloud/
6. Wendt, M.E.: Cloud-based hadoop deployments: benefits and considerations (2014). https://goo.gl/re0Ov5
7. OpenStack: Openstack sahara user documentation (2016). http://docs.openstack.org/developer/sahara/userdoc/overview.html
8. Mirantis: Openstack sahara kilo images (2016). http://sahara-files.mirantis.com/images/upstream/kilo/
9. Cloudera, I.: Cloudera manager free edition user guide (2012)
10. TPC: Dell poweredge r720xd with vmware vsphere 6.0 (2015). http://www.tpc.org/5504
11. OpenStack: Install and configure a storage node - openstack kilo (2015). http://docs.openstack.org/kilo/install-guide/install/yum/content/cinder-install-storage-node.html
12. RedHat: Cpu pinning and numa topology awareness in openstack compute (2015). http://redhatstackblog.redhat.com/2015/05/05/cpu-pinning-and-numa-topology-awareness-in-openstack-compute/
13. OpenStack: Openstack cinder multi-backend (2015). https://wiki.openstack.org/wiki/Cinder-multi-backend

From BigBench to TPCx-BB: Standardization of a Big Data Benchmark

Paul Cao[1], Bhaskar Gowda[2], Seetha Lakshmi[3],
Chinmayi Narasimhadevara[4], Patrick Nguyen[5], John Poelman[6],
Meikel Poess[7], and Tilmann Rabl[8,9(✉)]

[1] Hewlett Packard Enterprise, Palo Alto, USA
[2] Intel Corporation, Hillsboro, USA
[3] Actian Corporation, Palo Alto, USA
[4] Cisco Systems Inc., San Jose, USA
[5] Microsoft Corporation, Redmond, USA
[6] IBM, San Jose, USA
[7] Oracle Corporation, Redwood City, USA
[8] Technische Universität Berlin, Berlin, Germany
rabl@tu-berlin.de
[9] DFKI GmbH, Berlin, Germany

Abstract. With the increased adoption of Hadoop-based big data systems for the analysis of large volume and variety of data, an effective and common benchmark for big data deployments is needed. There have been a number of proposals from industry and academia to address this challenge. While most either have basic workloads (e.g. word counting), or port existing benchmarks to big data systems (e.g. TPC-H or TPC-DS), some are specifically designed for big data challenges. The most comprehensive proposal among these is the BigBench benchmark, recently standardized by the Transaction Processing Performance Council as TPCx-BB. In this paper, we discuss the progress made since the original BigBench proposal to the standardized TPCx-BB. In addition, we will share the thought process went into creating the specification, challenges in navigating the uncharted territories of a complex benchmark for a fast moving technology domain, and analyze the functionality of the benchmark suite on different Hadoop- and non-Hadoop-based big data engines. We will provide insights on the first official result of TPCx-BB and finally discuss, in brief, other relevant and fast growing big data analytic use cases to be addressed in future big data benchmarks.

1 Introduction

Organizations are increasingly beginning to value big data analytics for improving business, reducing the risks, and solving business challenges. At the same time, they are faced with a number of big data technology and solution options such as: MapReduce, Spark, NoSQL databases, SQL on Hadoop databases, and Flink. Choosing the right technology (or set of technologies) is critical for their success. A standardized benchmark that can be used to evaluate the performance of different big data technologies can greatly help organizations choose the right solution.

© Springer International Publishing AG 2017
R. Nambiar and M. Poess (Eds.): TPCTC 2016, LNCS 10080, pp. 24–44, 2017.
DOI: 10.1007/978-3-319-54334-5_3

Influenced by Moore's law, the rapidly evolving computing and storage landscape enables companies to analyze their data for half the cost every two years. Many companies hope to improve their business model by collecting increasing amounts of data and employing techniques related to big data. Although traditional database systems provide means to store large amounts of data, these have to generally need be in a structured format. In recent years, a large ecosystem of big data tools has evolved, which is targeted at analyzing the growing amounts of data, structured, semi-structured, or un-structured.

While database systems are well established and their performance is understood by companies, there is no easy methodology to compare, the plethora of big data systems with their many interfaces, APIs, and query languages. In certain situations a scalable big data system can be outperformed by a laptop for real problem sizes [1], emphasizing the need to improve efficiency of scalable big data systems.

Trying to keep up with this rapidly moving trend, customers have the difficult task on their hands to compare cross-platform solutions in order to select the right hardware and software for their big data needs. They rely on industry standard benchmarks to educate, inform and guide making these decisions. An absence of such performance analysis tools in form of standardized benchmarks has magnified customer difficulties, thus motivating the industry to take necessary actions to fill the void.

BigBench [2] was proposed to fill this gap, it set in motion efforts to create an end to end benchmark for big data analytics systems. While it comes with a concrete default implementation, the rules are very flexible regarding the type of systems this work can be run on and how the workloads can be implemented.

Thanks to member companies in the benchmark sub-committee under Transaction Processing Council (TPC), who contributed significant effort in drafting the specification and provide a readily usable benchmark kit, TPCx-BB progressed from being a scientific proposal [2] to an industry standard big data analytics benchmark in a span of two and half years.

In this paper, we describe the process towards a standardized benchmark and show how this process worked for BigBench. In particular, we have the following contributions:

- We give a detailed update of the benchmark and the changes that we required for the standardization.
- We present the first official benchmark submission and give an analysis on the results.
- We give an overview of existing BigBench implementations and compare them based on completeness.

The rest of the paper is structured as follows. In the next section, we give a brief overview of different big data benchmarking proposals. In Sect. 3, we present TPCx-BB and in Sect. 4, we describe its standardization process. Section 5 presents TPCx-BB experiments using different big data frameworks. Section 6 gives an outlook on future big data benchmarks and workloads. Section 7 concludes the paper.

2 Related Work

While several benchmarks for big data systems have been proposed, and discussed, most of them are either simplistic (e.g., limited to sorting or counting) or collections of simple use cases rather than end-to-end, application-level benchmarks. While these component benchmarks are good to test individual parts of a big data system, they cannot provide a holistic view of the performance of the system under test. And more importantly, none of these benchmarks have been discussed and reviewed under the umbrella of benchmark standardization organizations.

The Transaction Processing Performance Council[1] (TPC) understood this need and worked on several benchmarks for the big data space. As a stop-gap solution for MapReduce systems, TeraSort was standardized in TPCx-HS [3]. It is capable of indicating the basic I/O and network throughput of a MapReduce deployment but has limited other information value. Another ongoing work is the revision of TPC-DS [4] for big data systems. To this end, TPC-DS was adapted in Version 2 to accommodate the limitations of current "SQL on Hadoop" systems such has Apache Hive, Apache SparkSQL, and Apache Impala.

3 TPCx-BigBench (TPCx-BB)

Prior workshops on big data benchmarking have concluded that for successful adoption, a benchmark should have some relevance to their use cases, simple to implement, and easy to execute [5]. The TPC has a track record of publishing valuable and widely adopted benchmarks for measuring the performance of database systems. TPC-C, TPC-H, and TPC-DS are noteworthy enterprise benchmarks. Recently the TPC provides another option called as 'TPC Express' standard. Express benchmarks provide ready to run workloads to be executed on specific products. Here workload is bundled in the form of benchmark kits that are ready to run on a number of pre-selected platforms. The express benchmark model is very promising as it will lower the entry cost for test sponsors publishing the benchmark results. However, commitment of resources is required from the kit sponsor to develop, maintain, support and ratify the kit with in the sub-committee, for the lifetime of the kit. In designing the benchmark for big data systems, the TPC applied the lessons distilled from the making of previous successful and not so successful benchmark specifications. For example, with over 250 audited results publications and an even a larger number of publications that had used the benchmark to quantify and demonstrate performance gains from specific HW/SW enhancements, TPC-H is a widely successful benchmark, even though it has been criticized for not being representative of real world decision support workloads at high scale factors. In contrast, there has not been a single audited results published for TPC-DS benchmark, a richer and more comprehensive decision support benchmark, addressing the deficiencies in TPC-H and has been available since 2006. The success and popularity of TPC-H can be attributed to its relative simplicity (8 tables and

---

[1] Transaction Processing Performance Council – www.tpc.org.

22 queries) and timeliness when the database industry was making rapid advances in the data warehousing space and was in need of a relevant benchmark. On the other hand, with TPC-DS, it is a daunting task for end users to comprehend all the 99 queries, the rules for data refresh, the complex business problems designed to model and to analyze their performance. There have been some research publications or competitive analysis using only a subset (or modified versions) of the TPC-DS queries [6].

Balancing the thoroughness of an enterprise benchmark with the flexibility of an express benchmark while keeping the benchmark complexity under check, the TPCx-BB [7] took a middle of the road approach, in that it limited the number of queries to 30. To keep the benchmark relevant for the big data analytics use cases, the 30 queries are distributed to operate on structured, semi-structured, or unstructured data and using pure HIVE queries, MapReduce, natural language processing, or machine learning libraries. Further, to promote easy and quick adoption of the benchmark, a self-contained kit of the TPCx-BB is made freely available for download from the TPC website[2]. This kit can be used to measure the performance of Hadoop based systems including MapReduce, Apache Hive, and Apache Spark Machine Learning Library (MLlib).

3.1 TPCx-BB Overview

TPCx-BB is a big data batch analytics benchmark inspired by TPC-DS. The benchmark which models aspects of commercial decision support systems for a retail business. TPC-DS consists a snowflake schema representing three sales channels, (store, web, catalog, and online. Each with a sales and a returns table) and inventory fact table. The TPCx-BB uses the store and online distribution channels of TPC-DS and augments it with semi-structured and unstructured data. The prototype proposal of TPCx-BB was been discussed in detail [8].

3.2 Benchmark Kit

The kit is the first application-level benchmark suite specifically designed to measure the performance of big data analytics systems. TPCx-BB measures the performance of Hadoop-based systems including MapReduce, Apache Hive, and Apache Spark and its machine learning library MLlib, and is publicly available for download as a self-contained kit via the TPC Web site.

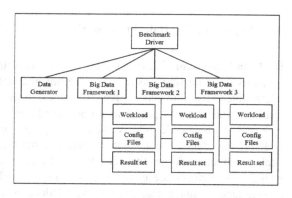

Fig. 1. Benchmark kit

---

[2] http://www.tpc.org/tpcx-bb.

TPCx-BB's benchmark kit is self-contained to have minimal requirements on external software dependencies and able to run 'out of the box' on the system under test (SUT). The kit is modular and it supports extensibility to new frameworks (i.e. collection of Big Data software/hardware components) can be easily added. The kit consists of three major components as shown in Fig. 1, (i) the benchmark driver, (ii) the workload (iii) the data generator.

Benchmark Driver. Implemented using Java and Bash scripts, the versatile benchmark driver is the heart of the kit. It orchestrates the workflow involved in executing the benchmark on the SUT. Support for running multiple concurrent query streams, automated answer set validation, SUT configuration details, and computing the benchmark score are done seamlessly at various phases during the benchmark execution. Additionally, the driver exposes hooks for integrating new frameworks as needed. An advanced mode the benchmark driver provides options to run the complete benchmark or individual queries for testing and optimization purposes.

Data Generator. The kit includes a parallel data generator based on the Parallel Data Generation Framework [9] to generate the input data set required for the benchmark. It is implemented as a Java program that runs as a MapReduce job on the SUT and can generate hundreds of terabytes of data in a relatively short time.

Workload. The kit is designed to have self-contained modules for each framework capable of running the TPCx-BB. All necessary binaries, configuration files, and answer set reside inside the framework module. This makes it easy for kit maintenance and help minimize the impact of adding new frameworks on existing kit modules. Addressing the complexity of big data frameworks and understanding the need to tune and optimize the benchmark, various configuration files provide sufficient hooks to tune the full benchmark or each individual queries by passing run time optimization parameters. Spark machine learning library suite is used for those queries invoking machine learning stages. OpenNLP framework is packaged with the kit for procedural programs invoking natural language processing.

3.3 Supported Big Data Frameworks

Big Data Ecosystem. Big data has transformed industries and research, spawning new solutions for addressing a wide range of technical challenges. Big data ecosystem today offers different end-to-end analytic strategies, scale-up frameworks for operational analytics, and scale-out platforms for advanced analytics.

Scale-up frameworks offer vertically integrated analytical workflows for medium scale big data datasets, e.g. database, data warehousing and online analytical systems. Scale-out frameworks on the other hand offer an array of frameworks closely mimicking high performance computing systems for analytics workflows requiring processing large complex datasets, e.g., MapReduce, Spark.

There are a number of execution frameworks that are part of the Hadoop ecosystem, including MapReduce, Spark, Tez, Flink, Storm, and Samza, each with its own strengths and weaknesses. Initially Hadoop was developed as a special-purpose infrastructure for big data with MapReduce handling massive scalability across hundreds or thousands of servers in a cluster. A number of vendors have developed their own distributions, adding new functionality or improving the code base derived from the Apache open source community. The most popular of these distributions are Cloudera, Hortonworks, MapR and IBM BigInsights each with their unique set of offerings.

SQL on Hadoop. One of the three V's used to describe Big Data is "Variety." Despite the diversity of data stored in Big Data systems, much of it still structured or can be transformed into a form with enough structure that a broad range of useful queries can be expressed in SQL. Evidence that SQL is still popular in the big data space can be seen in the plethora of SQL on Hadoop offerings available today. Some of these SQL engines for big data were built from the ground up to address big data problems, but many have a much longer history. For example, traditional database vendors including Oracle, Teradata and IBM have come out with versions of their SQL engines that run on Hadoop clusters.

One of the earliest and perhaps the most widely known SQL on Hadoop engines is Apache Hive. Hive supports a SQL-like language called HiveQL. Hive can execute queries using MapReduce2, Tez, or Spark. The TPCx-BB kit supports execution of the benchmark using Hive in all three of these frameworks. Besides Hive, there are several other SQL engines in open source, such as Apache Drill, Apache Phoenix, SparkSQL, Cloudera Impala, Teradata Presto, and Pivotal Hawq. Work is being done to have SparkSQL to fully support TPCx-BB, at the time of writing this paper, SparkSQL with help of support patches can successfully run all 30 queries. With the release of Spark 2.0, it is expected TPCx-BB should be able run on SparkSQL with no additional patches.

Non Hadoop Frameworks. TPCx-BB is a good fit for engines designed for processing or aggregating large amounts of data and that can either natively execute the machine learning and natural language processing required by BigBench, or can call out to other engines or frameworks such as Spark.

Since TPCx-BB kit has a pluggable architecture, support for additional SQL engines can be added over time. In fact, any engine capable of answering the 30 BigBench queries is a candidate for inclusion in the kit. The query syntax used by a given engine does not matter, since TPCx-BB allows the 30 use cases to be expressed in any SQL-like query language or natively written programs. However, since the queries are already expressible and available in HiveQL, developing implementations for SQL over Hadoop engines is usually straight forward and less involved than for engines whose query syntax is not similar to SQL. The benchmark prototype was implemented on two non-Hadoop frameworks, namely Apache Flink and Metanautix. As a matter of fact, the first BigBench prototype was actually implemented in Teradata Aster SQLMR.

Apache Flink is a big data streaming dataflow processing engine compatible to the Hadoop stack. It is based on the Stratosphere project [10]. Flink combines MapReduce functionality (e.g., schema flexibility and rich user defined functions) with techniques from traditional relational database management (e.g., query optimization, custom memory management, and pipelined processing) and adds dataflow and iterations. While having a different architecture, it offers similar functionality as Apache Spark and is, therefore, a candidate for a comparative benchmark implementation.

Quest is a massively distributed query processing engine offering from Metanautix, part of Microsoft. Quest is fully ISO/ANSI SQL'99 compliant, with a several extensions. It natively supports document data structures The Quest engine also connects to many data sources and extends the industry-standard Parquet columnar format with statistics for faster processing. User-defined functions can be written in LUA, C#, Java, Python, or SQL. A SQL extension, called Pipelines, is used to group SQL statements for more complex processing, such as the Pearson correlation, or K-Means (see Appendix A).

Prototype implementations of the benchmark on Flink and Quest, proves TPCx-BB is capable of working on non-Hadoop frameworks. TPCx-BB are open to new implementations, where TPCx-BB can be used to compare the performance and scalability of big data offerings and drive innovation in this space.

TPCx-BB in the cloud. At the high level TPCx-BB does not differentiate running the benchmark on SUT hosted in a datacenter or in the cloud. In the case of Infrastructure as a Service (IaaS) offerings from various cloud vendors, the benchmark can run with right framework and version requirements are met. In the past, the benchmark was run in Amazon AWS using different Hadoop distributions. However, on Big Data as a Service (BDaaS) offerings where the big data framework is an integrated offering, the benchmark is yet to be tested, examples of such offerings are Amazon Elastic MapReduce and theDatabricksCloud. For a fully valid result, where a test sponsors uses TPCx-BB on BDaaS for results publication, it should be noted, that the benchmark mandates adherence to the TPC pricing specification. TPC is working on amending their pricing specification to include cloud based offerings and facilitate cloud based TPC benchmark publications.

4   TPC Standardization of Big Bench

Founded in 1988, TPC's goal is to create, manage and maintain a set of fair and comprehensive benchmarks that enable end-users and vendors to objectively evaluate system performance under well-defined, consistent and comparable workloads. Currently, the TPC offers six are enterprise benchmarks (TPC-C and TPC-E for OLTP, TPC-DI for data integration, TPC-H for data warehouse, TPC-VMS for virtualization and TPC-DS for big data) and three are express benchmarks (TPCx-V for virtualization, TPCx-HS and TPCx-BB for big data). The TPC offers in parallel to the above listed benchmark specification so called Common Benchmarks, i.e. TPC-Energy and TPC-Pricing. These benchmark standards guarantee that energy consumption and pricing is measured in a consistent way across all performance benchmarks.

One of the pillars on which the credibility of TPC benchmarks rests is its strict audit rules. Audit rules guarantee that each benchmark publication was done according to its specification. TPCx-BB result is certified either by an independent certified TPC auditor or a TPCx-BB pre-publication board. The method to use is under the discretion of test sponsor.

4.1   Challenges During the Standardization

Standardizing an industry standard, involves framing set of rule and governance models. The process of standardization is a complex, cumbersome and time consuming process even for Greenfield benchmarks. Furthermore, the complexity was increased in the case of TPCx-BB where the specification had to consider the existing benchmark prototype during the process. This entire process posed unique set of challenges for the TPCx-BB sub-committee. The sub-committee worked diligently to address each of these issues, reached consensus and finally voted unanimously to launch benchmark. In this section, we make an attempt to present few selected challenges occurred during the standardization process, addressing previously uncharted areas in any TPC specification.

Execution Rules. The benchmark specification defines a set of narrow rules to ensure the results are consistent with the standard, auditable by an independent auditor and close any potential for gaps, which could be exploited to create benchmark specials. In TPCx-BB run rules requires the benchmark to be run two times for performance and repeatability of the results. The lower (i.e., worse) result metric of the two runs is reported. Each run must include, Data generation, load test, power test, throughput test and result check. The benchmark also adds an additional test to validate the query answer set for consistency by running scale factor 1 on the SUT. The results along with supporting files are audited for correctness by a TPC auditor or the publication board before publishing the result. The sub-committee spent considerable time in providing various tuning, and optimization options for test sponsors to experiment and get the best results possible, without breaking any of the rules. In addition to tuning the framework, the benchmark kit provides run time tuning options at global level where the tuning parameters are applied for the benchmark as whole and tuning individual queries by passing explicit parameters for a query. The benchmark specification provides clearly defined areas with examples in the appendix for such tunings. In an effort to keep answer sets for consistent for engine validation test, the sub-committee has put in place a set of rules to accommodate the differences between various query engines. This helps not only addition of future frameworks, but also fast evolving SQL on Hadoop frameworks like Hive. The benchmark also applies TPC-Pricing specification where necessary, which is mandatory for published results and provides the option to report the TPC-Energy metric.

Scale Factor. TPCx-BB's data set scales linearly with the scaling factor (SF). In order to be realistic across a large bandwidth of data set sizes (1 GB to 1 PB), the individual tables do scale in different ratios. While the large fact tables (sales and returns) scale linearly, other tables scale logarithmic or are completely static. Although this is realistic, it means that the ratio of sizes of the table changes with scale factors, e.g., for SF 1 the ratio of fact tables to dimension tables is approximately 50:50, while for large SFs the ratio becomes shifted to the fact tables. While BigBench scales continuously, TPCx-BB only specifies specific scale factors similar to TPC-H and TPC-DS (1, 3, 10, 30 …). Minor adjustments were made to the individual table scaling to ensure very close to linear scaling behavior for the full data set.

Metric. TPCx-BB's metric underwent a series of changes along with the execution model until its final version made it to the standard. The initially proposed metric was specified as the geometric mean of the execution time:

$$BB = \sqrt[4]{T_L * T_D * T_P * T_B} \tag{1}$$

where T_L is the time taken for loading the data into the system, T_D is the time for declarative queries, T_P is the time taken to process all procedural queries, and T_B is the time to process mixed queries. The query type is based on the implementation of the queries (declarative, procedural, or both).

However, since this is different for different kind of systems an alternative metric was proposed.

$$BB = \sqrt[30]{\prod_{i=1}^{30} P_i} \tag{2}$$

which also uses the geometric mean, but rather than summing the queries according to the classes uses each processing time individually. Both metrics only consider a power test style setup, where each query is processed individually and do not account for multi stream setups, where multiple users submit queries to a system. Also, they measure the runtime directly, meaning a smaller result is better. To improve this, a new metric was proposed in [20], which changed from a geometric mean to an arithmetic mean for all parts and incorporated not only the stream use case (throughput test *TT*) but also a data maintenance step (*DM*). The metric is scaled by the number of streams (*S*) to compute the total number of queries processed per hour (3600 s) incorporating regular updates (individual times are measured in seconds):

$$BBQph = \frac{30 * 3 * S * 3600}{S * T_L + S * T_P + T_{TT_1} + S * T_D + T_{TT_2}} \tag{3}$$

Although easy to understand, the arithmetic mean is not ideal in the case of highly skewed processing times. Since some queries process much less data than others and the data size processed does not scale linearly with the scaling factor for all queries, this is an issue in TPCx-BB. In this case, some queries will have very limited influence on the result of the metric. Therefore, a combination of geometric mean and arithmetic mean was finally incorporated in the standard:

$$BBQpm@SF = \frac{SF * 60 * M}{T_{LD} + \sqrt[2]{T_{PT} * T_{TT}}} \tag{4}$$

The load time T_{LD} (reduced by a factor of 10) is added to the geometric mean of the power test time T_{PT} and the throughput test time T_{TT}. Again, all times are measured in seconds but the metric is reported per minute (60 s). The number of queries (M) is divided by the sum of load and processing time, in order to get larger results for larger scale factors, the metric is multiplied by the scale factor (SF). While the power test time is compute as the geometric mean of all individual query processing times, the throughput time is the total processing time of all streams divided by the number of streams. Although not as easy to understand as the second metric, the final metric finds a good compromise for enabling useful optimizations.

Machine learning techniques. Three queries in TPCx-BB implement clustering, regression, and classification at various stages to satisfy the use case requirements. The benchmark kit uses algorithms bundled with Apache MLlib to invoke machine learning stages. Differing from standard based SQL API's where answer sets can be matched with relative accuracy, in machine learning techniques it is expected to see changes in answer set for two reasons, (a) changes to the algorithm in the same machine learning library for different versions, (b) introduction of a new machine learning library which may use a different method to implement an algorithm. TPCx-BB being an end to end system performance benchmark, leaves validating accuracy of an algorithm outside the scope of the specification. However, foreseeing these issues the specification provides general guidelines to address answer set changes triggered by change in library versions. In case no other changes apart from library updates are in the code or parameters in the benchmark kit the results are consider as valid and the reference results can be updated. In the case of new machine learning library, the new implementation may modify the code and parameters in the benchmark kit, but needs to use the same input data set and needs to match or improve the algorithm accuracy provided in the existing library. TPCx-BB addresses these variations in the specification of machine learning for the first time and thus, eases extensions of the benchmark and integration of changes during the lifecycle of the benchmark.

Determinism Requirements. SQL queries written for benchmarks are typically reproducible. They always return the exact same result independent of the execution engine. This is an important requirement for auditing since it enables verifying the correctness of query results and ensures all SUTs actually have to perform the same work. TPCx-BB contains several non-SQL workloads, some of which are machine learning tasks. These are typically implemented in a non-deterministic way and different algorithms can produce different results. In fact, the result quality typically depends on the number of iterations an algorithm has run for (up to the maximum achievable quality for an algorithm). This is a challenge for performance benchmarking, since result quality can be traded for performance. To alleviate this problem the kits algorithms are designed in a way that they produce the exact same results, or – where this is not possible – other implementations' algorithm have to have at least the same quality as the default implementation.

Reaching consensus. Although BigBench was fully implemented in a kit when it was proposed to the TPC, the specification had to be extended to cover all required regulations and rules. In this process, multiple changes were introduced to, one the one hand, fix minor deficiencies and to, on the other hand, not penalize certain vendors that have slightly different/not completely compatible functionality. This is one of the most delicate parts of standardization, since disagreement on this level can delay or even stop a benchmark standardization. One of the more controversial topics during the standardization of BigBench was the metric, as briefly touch upon above. To solve this, the

TPC subcommittee went through the process of preparing a model that can estimate performance, based on previously collected information, and using this to estimate the result of an execution. Being able to rethink and discuss setups with some numbers rather than on a theoretical level made it much easier for the committee to reach consensus.

5   Experiments with TPCx-BB Benchmark

In this section, we present experiments that were executed on independent test platforms, different frameworks, and small and large scale factors. We also discuss the hardware resource utilization behavior of one of the test platforms. Table 1 shows test details of the experiments.

The test runs were conducted with default settings, except where parameters needed to be configured to ensure all queries are able to run successfully. The dataset was generated using the default data generator and the tests were run using the driver provided in the kit.

Table 1. Test run experiments

Test #	Nodes in cluster	Framework	Scale factor
1	9	Hive on MapReduce	3000
2	8	Hive on Spark	1000
3	8	Hive on Tez	3000
4	8	SparkSQL	3000
5	1	Metanautix	1
6	8	Apache Flink	300
7	60	Hive on MapReduce	100000

5.1   Experimental Results

Test 1. The original implementation of the benchmark uses *Hive on MapReduce*. The test platform was configured with suitable parameters for Yarn, HDFS, and Hive, the benchmark was run with all three phases with two concurrent streams (default value) and completed successfully. Phase elapsed times were: load: 2803 s, power: 34076 s, and throughput: 54705 s.

Test 2. *Hive on Spark* utilizes Apache Spark as execution engine for Hive. Hive on Spark reuses Hive's planner/optimizer. The primary benefit is that Hive on Spark automatically gets full compatibility with all of Hive's features. The benchmark can run with Hive on Spark, with small changes in the configuration and changes on the cluster to enable Hive to use Spark as the execution engine. All the three phases of the benchmark completed successfully on the test platform. Phase elapsed times are: load: 9389 s, power: 13775 s, and throughput: 13864 s.

Test 3. *Tez* is designed to run batch and interactive workloads using the Hive API. In this test the load phase completed successfully, in the power phase 29 of 30 queries completed successfully. However Q16 failed to complete throwing an exception. The elapsed times for load was 3719 s.

Test 4. *SparkSQL* is an offering from Apache Spark to process structured data. SparkSQL is compatible with Hive, making it possible to run queries written in HiveQL without modifications. Enabling SparkSQL support for all 30 queries has been a multi month effort, where the benchmark team worked with the Apache Spark community to identify and fix missing features and bugs that prevented the complete execution of TPCx-BB queries. In this test, we had to apply a patch to Spark version 1.6.1 to get all queries to run successfully. This patch should be made available in yet to release Spark version 2.0. All three phases of the benchmark completed successfully on the test platform. Phase elapsed times were: load: 7896 s, power: 24,228 s, and throughput: 40,352 s.

Test 5. The *Metanautix* query processing engine is part of Microsoft's big data portfolio. All of the TPCx-BB queries were translated in SQL including sentiment analysis using a combination of window functions, user-defined Java functions, and pipelines. The machine learning post-processing stages were excluded.

Test 6. *Apache Flink* is a big data streaming dataflow processing engine compatible to the Hadoop stack. While having a different architecture with a purely stream-oriented execution engine, it offers similar functionality as Apache Spark. As a proof of concept, 22 queries were implemented using Flink'sDataSet API. In order to cover all necessary machine learning capabilities, a Flink-backed SystemML implementation was used for two of the queries [11].

Test 7. The objective of this test to demonstrate readiness of the benchmark to scale beyond small dataset and clusters. For this purpose, we selected a cluster with 60 nodes and dataset scale factor of 100000 which is close to 100 TB of input data. Hive on MapReduce was used as execution framework. We ran load and power phase and skipped the throughput phase due to limited availability of cluster time. Phase elapsed times were: load: 19,941 s and power: 401,738 s. During the tests, we found that the usage of realistic data distribution models in the benchmark result in a number of skewed tasks on Hive on MapReduce, where skewed tasks processes many more records than others and took much more time to complete. While this behavior is seen across all scale factors and cluster sizes, the result is amplified running the benchmark on the larger dataset and more number of nodes, challenging the efficiency of the query engine.

This set of experiments shows that various big data frameworks are able to run the benchmark with modification or no modifications, as demonstrated by experiments 1–6. This proves the versatility of the benchmark kit and shows that it can be used to compare and distinguish multiple frameworks for their features and performance. Partial execution of the kit on Metanautix shows that non-Hadoop-based frameworks are capable of adapting the benchmark. Partial execution of the benchmark on Apache Flink demonstrates the system agnostic nature of TPCx-BB, the use cases can be implemented natively without higher level SQL expression API's. Data and cluster scale tests bring out issues which are mostly uncaught during the development stages proving that a benchmark's role goes beyond providing publications but also helping vendors iteratively tune their platforms.

5.2 Resource Utilization Tests

Hardware platform tuning is often used to optimize the SUT to its maximum efficient state, i.e., the configuration where the test hardware is fully utilized with no obvious bottlenecks. Analysis of the hardware behavior under the load is crucial to understand the baseline performance and identify and resolve any bottlenecks. In this section, we analyze hardware resource utilization comparing the utilization patterns of the test platform by running the benchmark two times on a fixed hardware setup, scale factor, and big data framework. In the second test, we increase the number of concurrent streams in the throughput phase from 2 to 4.

Benchmark Setup. The cluster consists of eight HPE DL360 G8 nodes, with the configuration shown in Table 2. The experiments were conducted running all three phases of TPCx-BB on Scale Factor 3000. Hive on MapReduce was selected as the framework. Intel's Performance Analysis Tool[3] was used to collect the utilization pattern from the cluster nodes.

Table 2. Cluster configuration

Node	Role	Hardware	Software
1	Master server	24C, 192 GB RAM, 8.5 TB storage, 10 Gbe	RHEL 6.7, CDH 5.6
2–8	Worker node	24C, 256 GB RAM, 8.5 TB storage, 10 Gbe	RHEL 6.7, CDH 5.6

Table 3 shows the elapsed times for load, power, and throughput phase for both of the test runs. The load phase consists of reading the generated data to create the test dataset in appropriate format; copy data into final location; data preparation including metadata creation, population, and computation of database statistics. The power phase is designed to measure the performance of the SUT when processing all the queries in sequential order. The elapsed times for load and power phases are comparable with variation expected from a Hadoop system. In this test we are in particular interested in the system characteristics of the throughput phase. During this phase, all queries are executed using concurrent streams. Each query stream runs all queries, where each stream has a different order of queries. As can be

Table 3. Elapsed times

Phase	2 Streams	4 Streams
Load	2803	2796
Power	34076	34179
Throughput	54705	104565

---

[3] PAT - https://github.com/intel-hadoop/PAT.

seen in the table, the elapsed time for the throughput phase doubles for 4 concurrent streams in comparison to 2 concurrent streams.

Analysis of Utilization Pattern. The charts in Fig. 2 show the hardware utilization pattern behavior of the cluster when running the benchmark with 2 and 4 streams. The chart shows comparison of the major components of the cluster, i.e., CPU utilization, memory utilization, I/O bandwidth, and network I/O. Since we have captured data at one second samples, the chart is compressed on the time scale to show the complete execution of the benchmark.

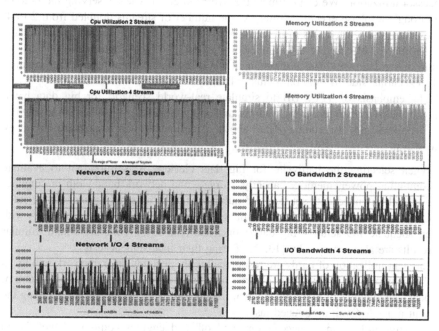

Fig. 2. Processor, memory and I/O utilization

The first mark in the time scale in Fig. 2 marks the end of the load phase of both test which is at ∼2,800 s. The load phase involves data staging and replicating over data nodes that results in a cluster management overhead. This governs the performance of this stage with significant CPU and memory utilization, I/O bandwidth, and network I/O. The load phase uses software compression to compress the raw input data into optimized columnar format, resulting in additional CPU utilization.

The power phase utilization can be seen between the first and second mark in the time scale in Fig. 2. The individual peaks are signatures of each query being run in sequential order. Additional insight of the queries can be gained by mapping the running time of each to the time dimension on the charts. The independent utilization pattern for each query highlights that, unlike the constant ramp-up and down seen in micro-benchmarks, TPCx-BB exhibits use case driven utilization patterns close to real world big data use cases, where the platform needs to accommodate both short and long running tasks. It is possible to go into more fine granular analysis of each query and gain insight into the individual system resource usage. This leads to a better understanding of the query and system behavior when tuning individual queries. As expected the power phase shows very similar comparative elapsed times between the two tests.

In Fig. 2, the second mark indicates the start of the throughput phase, the CPU utilization shows a steady high processor usage. A more detailed analysis showed 70% utilization for two concurrent streams and 90% utilization for four concurrent streams. The memory, storage, and network I/O are sufficiently utilized but nowhere close to the processor utilization. We can estimate the overhead effect when observing the ratio of the throughput phase execution time. As the number of streams doubled from 2 to 4, the execution time increases by a factor of 2. The overhead of running more streams can be inferred by varying the number of streams. The throughput phase reflects the nature of big data workloads comprising a mix of both short running and long running tasks executing side by side on a cluster [12].

The emphasis of TPCx-BB to simulate real-world scenario for big data batch analytics helps to extrapolate the findings and apply the takeaways when deploying big data applications. By running the above experiments we summarize few key takeaways:

- When selecting the hardware for big data clusters, it is important to evaluate computing power, memory capacity, storage, and network bandwidth in conjunction with intended data set size and number of tasks required to run side by side.
- Contrary to common belief that big data workloads are I/O bound, we notice – with an adequate I/O setup – big data workloads tend to be compute bound. Similar results are also reported by [13, 21] during their independent tests.
- Efficient utilization of hardware resources highly depends on framework tuning. In this example, we believe – as software schedulers evolve – the utilization pattern of peaks and valleys of will reduce when freeing hardware resources and reducing the wait times for waiting queued tasks.
- Selective utilization of accelerators and off-load engines could be beneficial to increase overall efficiency of the cluster. An example could be load phase compression off-load.

6 Benchmarking Emerging Big Data Use Cases

In recent years, there have been large advances in analytics software. As big data reaches a larger audience, the community has sought to commoditize general purpose algorithms and systems for increasingly elaborate analytical tasks. The generation of large datasets has been increasing, leading to the development of new big data processing frameworks, which is predominantly driven by "People and Things". For example, "People" interacting via social media portals and cloud enabled applications are driving an ever increasing volume of data into the cloud [14]. "Things" are intelligent and connected devices capable of making semi-autonomous decisions using models received by cloud-based or -hosted compute farms. Addressing these two important segments with a relevant benchmark, will help the industry and academic community to validate the performance of new implementations.

There is a broad range of new applications for these analytical capabilities, to name a few:

- Recommendation systems: graph processing, stream processing, machine learning.
- Search and ranking: graph processing, machine learning.
- Fraud detection: machine learning, *ad-hoc* analysis.
- Internet-of-Things (IoT): stream processing, lambda processing.
- Image, video, audio, and natural language processing: deep learning using neural networks.

For the purpose of this paper and benchmarking, we select two categories of the advances as follows:

- Processing frameworks
- Machine learning

6.1   Processing Frameworks

Stream. Stream processing is mainly used in real-time analytics, where the events are streamed in form of micro or mini batches. A data stream can be as simple as time series events displayed in real-time, e.g., temperature readings from a sensor, or processed as complex events by applying computation techniques in real-time, e.g., identifying failed components in an airplane using anomaly detection techniques. In addition to acting on the incoming stream in real-time, events are stored for feedback-based learning and historical trend analysis using batch analytics.

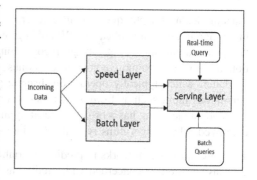

Fig. 3. Lambda architecture

Data from a device in the field can be permuted and aggregated at the source or in mid-way before it is transferred to the cloud. The lambda architecture [15] is an example of a stream processing framework using three layers of processing (Fig. 3):

- Batch Layer – curates the master dataset by storing all data entering the system using batch processing techniques.
- Serving Layer – enables fast *ad-hoc* insights extracted from data curated in the batch layer.
- Speed Layer – provides real-time insights from the incoming/streamed data, including running machine learning algorithms, on real-time data.

An IoT benchmark based on such an architecture can serve as an excellent proxy to test functions involved with streaming and real-time analytics.

Graph. Human interaction with the internet changed the Web 2.0 [16]. The emergence of various social networking platforms, search engine optimizations, and the ability to connect these human interactions with business models was unthinkable just a decade ago. Graph processing systems are used in analyzing networks of relationships normally represented in data objects referred as nodes and edges. Some large graph datasets can span trillions of edges [17].

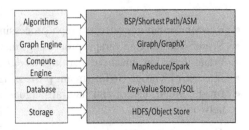

Fig. 4. Graph processing framework

Graph processing requires a robust framework with characteristics such as, fault tolerant storage, fast database, scale-out graph analysis engines, scale-out computation engine, and efficient algorithms as illustrated in Fig. 4.

6.2   Machine Learning

Machine learning techniques continue to grow in significance but also are expanding into different areas of application. With this growth the field is transitioning from a few "bespoke" applications; e.g., image recognition, machine translation, speech recognition, and robotics, to more commoditized ones; e.g., fraud detection. We will focus on the latter, which typically operate on discrete symbols such as words as opposed to continuous input such as from a microphone or historical revenue.

Machine learning has a broad range of applications with different algorithms being employed. These algorithms typically fall into two categories:

1. Regression, which works to predict a variable's value (e.g., projection of revenue),
2. Classification, is concerned with predicting a label for a sample (e.g. male/female, will or will not buy).

Moreover, a task can be structured where the prediction happens on a graph or sequence such as machine translation generating a sequence of words in a foreign language. The task can also be unstructured, where the desired output is a single value like the next stock price, or whether a fraud occurred or not. Training of a model can be supervised, unsupervised, or utilizing reinforcement learning. In each of these scenarios, one can define a measure of quality such as in the case of fraud detection;

1. A weighted sum of false positives - fraud was declared when a transaction was in good standing
2. False negatives - fraud remained undetected

Because the data are generated automatically, they have special properties which can be exploited by the algorithms. Therefore, as in TPCx-BB, we should factor in the speed of the algorithms.

TPCx-BB as a batch analytics benchmark provides excellent coverage for advanced analytics to examine large datasets. Most of the benchmark is implemented using data management primitives and functions. Although there are a handful of use cases in TPCx-BB invoking machine learning algorithms[4], TPCx-BB is far from being a comprehensive representation of analytics using machine learning algorithms. Currently, neither streaming processing, graph processing, nor deep learning are represented in TPCx-BB. Given the recent interest in deep learning, and its broad range of applicability, it should be given special consideration.

There have been some efforts in the analytics community to address these areas [18, 19]. However, there hasn't been any collaborative push from the industry and academia to create a use case based benchmarking framework. We think it would be impractical to expand the coverage of the TPCx-BB benchmark to include all of these, therefore, they should be the focus of future benchmarks.

7 Conclusion

In 2013, the proposal "BigBench" was brought to the attention of the analytics community as a candidate for a first end-to-end big data benchmark. Since then idea has evolved, been put under the scrutiny of experts and public alike to finally emerge as TPCx-BB, the first industry standard big data benchmark with relevance to big data use cases. During this process, several changes went into the benchmark, which we discussed in this paper. Preliminary results are encouraging and it already has seen adoption with first results being published[5]. The benchmark helps the big data software ecosystem to identify performance bottlenecks, feature gaps, and scaling issues, which previously often remained undiscovered. The benchmark has also helped driving innovation in non-Hadoop ecosystems.

In this paper, we have tracked the course of the BigBench journey, gave a snapshot of its current state and potential changes coming in the future. We have conducted extensive experiments using the benchmark, and offered observations and analyses of several platforms. This paper offers a glimpse of the TPC standardization process, challenges and means to navigate through them successfully.

Acknowledgements. We would like to thank Sreenivas Viswanada from Microsoft Corporation for running experiments on Metanautix. Yao Yi and Zhou Yi from Intel Corporation for their help to run 100 TB experiment. Michael Frank and Manuel Dansich from bankmark for their work on the TPCx-BB benchmark kit.

---

[4] Examples are clustering, logistic regression, and sentiment analysis.

[5] Hewlett Packard Enterprise ProLiant DL for Big Data – http://www.tpc.org/3501.

This work has been partially supported through grants by the German Ministry for Education and Research as Berlin Big Data Center BBDC (funding mark 01IS14013A) as well as through grants by the European Union's Horizon 2020 research and innovation program under grant agreement 688191.

Appendix A

K-Means using SQL. It is possible to write K-means using SQL and extensions in the Metanautix Quest system. The full implementation is complex, requiring an iteration (implemented using SQL triggers), but also rebalancing when a class becomes empty. For simplicity we assume that each point is described by an id, and a coordinate vector x. Using a SQL UDF, we can write the Distance function. A user-defined aggregation function, AVG_VECTOR, computes the average vector. We assume 50 classes. We outline the steps:

1. Initialization of class centroids

```
CREATE TABLE Centroids .. AS
  SELECT ROW_NUMBER() OVER (ORDER BY RANDOM()) r, x FROM
Data WHERE r <= 50
```

2. Assigning data points to classes

```
CREATE TABLE ClassAssignment .. AS
  SELECT id, r FROM Centroids C, Data D WHERE
Distance(D.x, C.x) = (SELECT MIN(Distance(D.x, C2.x))
FROM Centroids C2)
```

3. Compute new centroids

```
CREATE TABLE NewCentroids .. AS
  SELECT r, AVG_VECTOR(x) x FROM Centroids C,
ClassAssignment CA, Data D WHERE
  C.r = CA.r AND CA.id = D.id
```

Using window functions. Window functions can be used where a MapReduce, or multiple passes would be otherwise required. As an example, we show how Query 02 can be rewritten.

```
WITH Session as (
SELECT DISTINCT
sessionid,
wcs_item_sk
FROM
(SELECT
  *,
concat(cast(wcs_user_sk as string), '_', cast(bucket as
string)) sessionid
FROM
(SELECT
  *,
  (first(tstamp_inSec) over (partition by wcs_user_sk
                            order by tstamp_inSecdesc)
- tstamp_inSec) / 3600 bucket
FROM
  (SELECT
wcs_user_sk,
wcs_item_sk,
    (wcs_click_date_sk * 24 * 60 * 60 +
wcs_click_time_sk) AS tstamp_inSec
  FROM web_clickstreams
  WHERE wcs_item_sk IS NOT NULL
  AND   wcs_user_sk IS NOT NULL))))
```

References

1. McSherry, F., Isard, M., Murray, D.G.: Scalability! But at what COST? In: HotOS 2015 (2015)
2. Ghazal, A., Rabl, T., Hu, M., Raab, F., Poess, M., Crolotte, A., Jacobsen, H.-A.: BigBench: towards an industry standard benchmark for big data analytics. In: SIGMOD 2013 (2013)
3. Nambiar, R.O., Poess, M., Dey, A., Cao, P., Magdon-Ismail, T., Ren, D.Q.: Andrew bond: introducing TPCx-HS: the first industry standard for benchmarking big data systems. In: Nambiar, R., Poess, M. (eds.) TPCTC 2014. LNCS, vol. 8904, pp. 1–12. Springer, Cham (2014)
4. Poess, M., Nambiar, R.O., Walrath, D.: Why you should run TPC-DS: a workload analysis. In: VLDB 2007 (2007)
5. Baru, C., Bhandarkar, M., Nambiar, R., Poess, M., Rabl, T.: Setting the Direction for Big Data Benchmark Standards. In: Nambiar, R., Poess, M. (eds.) TPCTC 2012. LNCS, vol. 7755, pp. 197–208. Springer, Heidelberg (2013). doi:10.1007/978-3-642-36727-4_14
6. Ghat, D., Rorke, D., Kumar, D.: New SQL Benchmarks: Apache Impala (incubating) Uniquely Delivers Analytic Database Performance. https://blog.cloudera.com/blog/2016/02/new-sql-benchmarks-apache-impala-incubating-2-3-uniquely-delivers-analytic-database-performance/
7. Transaction Processing Performance Council. TPC Express Benchmark™ BB. http://www.tpc.org/tpcx-bb

8. Baru, C., Bhandarkar, M., Curino, C., Danisch, M., Frank, M., Gowda, B., Huang, J., Jacobsen, H.-A., Kumar, D., Nambiar, R., Poess, M., Raab, F., Rabl, T., Ravi, N., Sachs, K., Yi, L., Youn, C.: An analysis of the BigBench workload. In: TPCTC 2014 (2014)
9. Rabl, T., Frank, M., Sergieh, H.M., Kosch, H.: A data generator for cloud-scale benchmarking. In: Nambiar, R., Poess, M. (eds.) TPCTC 2010. LNCS, vol. 6417, pp. 41–56. Springer, Heidelberg (2011). doi:10.1007/978-3-642-18206-8_4
10. Alexandrov, A., Bergmann, R., Ewen, S., Freytag, J.-C., Hueske, F., Heise, A., Kao, O., Leich, M., Leser, U., Markl, V., Naumann, F., Peters, M., Rheinländer, A., Sax, M.J., Schelter, S., Höger, M., Tzoumas, K., Warneke, D.: The stratosphere platform for big data analytics. VLDB J. **23**(6), 939–964 (2014)
11. Boehm, M., Burdick, D., Evfimievski, A.V., Reinwald, B., Sen, P., Tatikonda, S., Tian, Y.: Compiling machine learning algorithms with SystemML. In: SoCC 2013 (2013)
12. Chen, Y., Ganapathi, A., Griffith, R., Katz, R.: The case for evaluating MapReduce performance using workload suites. In: MASCOTS 2011 (2011)
13. Ousterhout, K., Rasti, R., Ratnasamy, S., Shenker, S., Chun, B.-G.: Making sense of performance in data analytics frameworks. In: NSDI 2015 (2015)
14. O'Leary, D.E.: 'Big Data', the 'Internet of Things' and the 'Internet of Signs'. In: Intelligent Systems in Accounting, Finance and Management, vol. 20(1), pp. 53–65
15. Marz, N., Warren, J.: Big Data: Principles and Best Practices of Scalable Realtime Data Systems. Manning Publications, New York (2015)
16. Malewicz, G., Austern, M.H., Bik, A.J.C., Dehnert, J.C., Horn, I., Leiser, N., Czajkowski, G.: Pregel: a system for large-scale graph processing. In: SIGMOD 2010 (2010)
17. Ching, A., Edunov, S., Kabiljo, M., Logothetis, D., Muthukrishnan, S.: One trillion edges: graph processing at facebook-scale. PVLDB **8**(12), 1804–1815 (2015)
18. Li, M., Tan, J., Wang, Y., Zhang, L., Salapura, V.: SparkBench: a comprehensive benchmarking suite for in memory data analytic platform Spark. In: CF 2015 (2015)
19. Cooper, B.F., Silberstein, A., Tam, E., Ramakrishnan, R., Sears, R.: Benchmarking cloud serving systems with YCSB. In: SoCC 2010 (2010)
20. Rabl, T., Frank, M., Danisch, M., Gowda, B., Jacobsen, H.-A.: Towards a complete BigBench implementation. In: Rabl, T., Sachs, K., Poess, M., Baru, C., Jacobson, H.-A. (eds.) WBDB 2015. LNCS, vol. 8991, pp. 3–11. Springer, Heidelberg (2015). doi:10.1007/978-3-319-20233-4_1
21. Chen, Y., Choi, A., Kumar, D., Rorke, D., Rus, S., Ghat, D.: How Impala Scales for Business Intelligence: New Test Results. http://blog.cloudera.com/blog/2015/09/how-impala-scales-for-business-intelligence-new-test-results/

Benchmarking Spark Machine Learning Using BigBench

Sweta Singh[✉]

IBM, Dallas, USA
singhswe@us.ibm.com

Abstract. Databases such as dashDB are adding High Speed Connectors for Spark to efficiently extract large volumes of data. This allows them to be combined with other unstructured data sources and perform Machine Learning (ML) on top of it. Machine Learning is a key ingredient for such use cases. In order to assess performance of the data connectors and machine language frameworks, we sought benchmarks that have the ability to scale the size of datasets to very large volumes and apply Machine Learning algorithms. After exploring several options, we found BigBench to be a good fit. In this paper, we talk about our experiences of using BigBench with special focus on its 5 Machine Learning queries and their default implementation in Spark. We discuss on how we could improve effectiveness of BigBench for benchmarking Machine Learning by avoiding bias and inclusion of real time analytics. We also think that there is scope for improving the coverage of Machine Learning by adding more use cases like Collaborative Filtering. Lastly, we share some interesting visualization of 4 ML queries using SPSS Modeler and our experiments on different Clustering and Classification algorithms.

Keywords: Collaborative filtering using machine learning · Predicting accuracy of data sets · Visualization of bigbench machine learning queries using SPSS

1 Introduction

Machine Learning is being increasingly applied on large volumes of data to be able to predict outcomes with high efficiency and accuracy. The performance of such use cases are determined by two key aspects

(a) Optimized data exchange between analytics engines like Spark [1, 18, 19] and the data store. This is important due to the iterative nature of machine learning algorithms, especially for large data sets that do not fit in memory
(b) Scalability and accuracy of the Machine Learning frameworks

To address the first aspect above, several databases like IBM dashDB [2], Cassandra, Couchbase etc. are employing techniques for optimized data connectors that make bi-directional communication between Spark and databases more efficient by

(a) Generating parallel SQL statements under the covers, to read from multiple database partitions concurrently

© Springer International Publishing AG 2017
R. Nambiar and M. Poess (Eds.): TPCTC 2016, LNCS 10080, pp. 45–60, 2017.
DOI: 10.1007/978-3-319-54334-5_4

(b) Localizing the data exchange between Spark and database node if they are hosted on the same cluster

IBM dashDB Local [3] has the capability to run Spark applications that analyze data in a dashDB database and write results back to the database. It achieves highly optimized and parallel data transfer between dashDB and Spark by collocating DB2 data nodes and Spark executors. An I/O layer, implemented as a DB2 Fenced Mode Process (FMP), acts as an interface between dashDB and Spark. dashDB unloads the data of the local partition into the FMP, which then passes the data to the Spark layer. A single transfer unit contains multiple blocks and potentially millions of rows, hence providing significant speed up for exchange of data.

We sought benchmarks to evaluate the performance benefits of connectors. These were the key pre-requisites:

(a) The benchmark should be representative of a good use case for Spark and database integration. Machine Learning has to be a key component of the use case
(b) It should be able to scale data volumes, allowing for large data volume transfers and be bidirectional - read/write to database
(c) It should invoke Machine Learning algorithms via SQL interface (Stored Procedure) or via Spark jobs (using customized RDD to connect to data source)
(d) It should support multiple streams to test both scalability and resource management in an integrated solution where Spark and database co-exist on the same cluster

One option that we explored were open data sets available as part of UCI machine learning repository [4]. They have 336 datasets based on "real" world data, which is precisely the reason why they are an attractive option. However, after examining some of the new and popular data sets, we found that their key drawback was the small dataset size, ranging from few KB to less than 500 MB.

Given our key requirements of large data transfers, ability to scale and run multi-stream, we found BigBench [7–9, 17] to be an apt choice. There are five queries in BigBench (Q05, Q20, Q25, Q26 and Q28) that involve interaction between database and Spark. These cover three Machine Learning algorithms in Clustering (K-Means) and Classification (Logistic Regression and Naive Bayes). Since data exchange volumes will be large and data cannot be entirely cached in memory for reuse, the iterative nature of Machine Learning algorithms involves multiple trips to the data source and hence the performance of data connectors is stressed well enough. BigBench also supports bidirectional transfer between data layer and Spark by supporting writing back the scoring results to the database.

Our experiments were conducted with the following configuration:

BigBench Scale Factor = 1 TB
dashDB Local cluster, CentOS7.0-64 and Spark 1.6.2
4 nodes with the following configuration:
- 24 cores (2.6 GHz Intel Xeon-Haswell)
- 512 GB memory
- 6 internal SSDs (960 GB SanDisk CloudSpeed 1000 SSD)
- 10000 Mbps full duplex N/W card

We also studied the effectiveness of BigBench in assessing Machine Learning frameworks as we intend to extend our study to assess Machine Learning algorithm implementation in Spark MLlib versus IBM ML algorithms.

In the following sections, we discuss our key observations and recommendations.

2   Collaborative Filtering Using Machine Learning

We propose to add a "Recommender System" use case in BigBench. It is a very relevant use case in e-commerce and can be implemented using Collaborative filtering, one of the most widely used and researched methods. Collaborative Filtering is based on the principle that if two users rate x items similarly, they will likely rate other items similarly. It is known for two unique challenges:

(a) Data Sparsity: For a large customer base and a large product set, the subset of items rated by users is very small. This leads to a sparse user-item association matrix and hence poses a challenge to the predictions of the algorithm.
(b) Scalability: With a large customer base and range of items, the computational complexity of Collaborative Filtering grows very quickly. There are several optimizations to reduce the cost of calculating similarity but there is often a trade-off between performance and accuracy.

Matrix Factorization is one of the key methods to implement Collaborative Filtering. It is often classified as a latent factor model. The ratings are explained by characterizing both items and users on a number of factors automatically inferred from the ratings pattern [13–15]. The sparse rating matrix is modeled as the product of low-rank user and item factors. Latent factors are learnt by minimizing the reconstruction error of the observed ratings. The unknown ratings can then be computed by multiplying these factors.

Given that BigBench already provides the [user, product, review] triplet in web_clickstreams, Matrix Factorization will be a good add-on to the BigBench Machine Learning arsenal to study the performance of the Machine Learning framework and the cluster.

Spark MLlib [11] implements Matrix Factorization using Alternating Least Squares (ALS) method [16, 20–22]. ALS alternates between fixing the user feature matrix and the item feature matrix. When one is fixed, the other is solved by minimizing the Root Mean Squared Error. This is repeated until convergence.

Below, we share an initial prototype for BigBench Recommendation System using Explicit Feedback.

The input Vector can be extracted from product_reviews table in 3 possible ways

(a) Select all items, users, ratings from product_reviews table
    SELECT PR_USER_SK, PR_ITEM_SK, PR_REVIEW_RATING FROM product_reviews;
(b) Predict recommendations in a specific item category
    SELECT PR_USER_SK, PR_ITEM_SK, PR_REVIEW_RATING FROM product_reviews pr, item

WHERE I_CATEGORY = 'Books' AND i_item_sk = pr_item_sk
(c) Predict recommendations to specific item category and class
        SELECT PR_USER_SK, PR_ITEM_SK, PR_REVIEW_RATING FROM
        product_reviews pr, item WHERE I_CATEGORY = 'Books' AND I_CLASS =
        'fiction'

The Spark job for Recommender does the following:

(a) Accepts arguments for iterations and rank that are parameters of the ML algorithm
(b) Transforms Rows of data frame to Rating object
(c) Trains ALS model on a partial data set (90%) and predicts the outcome on the held back data
(d) Measures the accuracy using Root Mean Squared Error (RMSE)

Below is a snippet from Recommender output on 1 TB scale factor of BigBench. At this scale factor, the product_reviews table has 4.45 million ratings. 1.7 million customers have rated 556,000 items.

```
Actual Rating: 5.0 Predicted Rating: 4.46
Actual Rating: 5.0 Predicted Rating: 4.99
Actual Rating: 2.0 Predicted Rating: 1.99
Actual Rating: 5.0 Predicted Rating: 4.98
Actual Rating: 4.0 Predicted Rating: 3.95
Actual Rating: 5.0 Predicted Rating: 4.99
Actual Rating: 5.0 Predicted Rating: 5.42
Actual Rating: 5.0 Predicted Rating: 4.99
Actual Rating: 5.0 Predicted Rating: 4.99
Actual Rating: 5.0 Predicted Rating: 4.99
Actual Rating: 5.0 Predicted Rating: 4.99
Actual Rating: 5.0 Predicted Rating: 4.99
Actual Rating: 5.0 Predicted Rating: 4.99
Actual Rating: 4.0 Predicted Rating: 4.03
Actual Rating: 1.0 Predicted Rating: 0.94
Actual Rating: 5.0 Predicted Rating: 4.84
```

Spark uses a blocked implementation of ALS. A User column block stores a subset of user factor matrix and an Item column block stores a subset of item factor matrix. The rating matrix is stored as two separate RDDs. One RDD is partitioned by user and the other RDD is partitioned by item. While updating the user factor matrix, ratings for each user in its partition is locally available. However, item factor vector corresponding to items rated by the user must be shuffled across the nodes.

Below is a snapshot of ALS training stage execution with twenty iterations. Figure 1 shows that Job 3 is the most expensive job. Figure 2 shows a snippet of Job 3 breakdown. Job 3 comprises forty stages, two for each iteration. The user or the item factor matrix is updated alternately in each stage. At the end of each stage, the updated factor matrix is shuffled across the nodes. A major chunk of the time is spent in the shuffle.

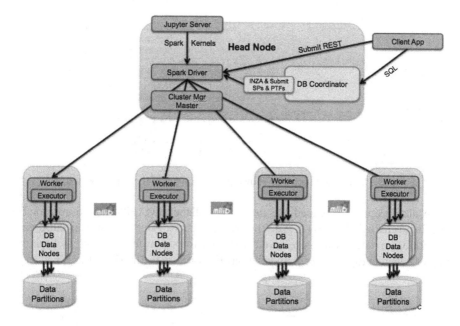

Fig. 1. dashDB Spark integration layout

Job Id	Description	Submitted	Duration
8	mean at Recommender.scala:167	2016/08/14 22:18:45	40 s
7	aggregate at MatrixFactorizationModel.scala:96	2016/08/14 22:18:45	0.5 s
6	first at MatrixFactorizationModel.scala:67	2016/08/14 22:18:45	14 ms
5	first at MatrixFactorizationModel.scala:67	2016/08/14 22:18:44	13 ms
4	count at ALS.scala:264	2016/08/14 22:18:44	0.8 s
3	count at ALS.scala:263	2016/08/14 22:12:47	5.9 min
2	count at ALS.scala:604	2016/08/14 22:12:43	3 s
1	count at ALS.scala:596	2016/08/14 22:12:33	10 s

Fig. 2. Spark monitoring UI showing the jobs

We did some experiments to study the impact of the number of factors or rank on the performance and accuracy of Collaborative Filtering. The intermediate RDDs are 100% cached. As we increase the rank, we observe that the accuracy of the algorithm changes, with consistent drop in performance. This performance drop is attributed to increase in shuffle time due to larger size of the factor matrix (Fig. 3).

Completed Stages (42)

Stage Id	Description		Submitted ▲	Duration	Tasks: Succeeded/Total	Input	Output	Shuffle Read	Shuffle Write
13	map at ALS.scala:752	+details	2016/08/26 06:33:25	7 s	192/192	42.5 MB			65.2 MB
14	flatMap at ALS.scala:1170	+details	2016/08/26 06:33:32	13 s	192/192	12.8 MB		65.2 MB	138.2 MB
	org.apache.spark.rdd.RDD.flatMap(RDD.scala:332) org.apache.spark.ml.recommendation.ALS$.org$apache$spark$ml$recommendation$ALS$$ computeFactors(ALS.scala:1170) org.apache.spark.ml.recommendation.ALS$$anonfun$train$1.apply$mcVI$sp(ALS.scala: 643)								
15	flatMap at ALS.scala:1170	+details	2016/08/26 06:33:45	9 s	192/192	45.0 MB		138.2 MB	110.4 MB
16	flatMap at ALS.scala:1170	+details	2016/08/26 06:33:53	8 s	192/192	55.3 MB		110.4 MB	131.7 MB
17	flatMap at ALS.scala:1170	+details	2016/08/26 06:34:01	8 s	192/192	45.0 MB		131.7 MB	110.4 MB
18	flatMap at ALS.scala:1170	+details	2016/08/26 06:34:10	8 s	192/192	55.3 MB		110.4 MB	131.7 MB
19	flatMap at ALS.scala:1170	+details	2016/08/26 06:34:18	9 s	192/192	45.0 MB		131.7 MB	110.4 MB
20	flatMap at ALS.scala:1170	+details	2016/08/26 06:34:27	8 s	192/192	55.3 MB		110.4 MB	131.6 MB

Fig. 3. Spark monitoring UI showing a breakdown of the expensive job

We also propose to extend the usage of Collaborative Filtering to construct a streaming scenario in BigBench. We can stream a portion of web click stream data during the workload execution and use saved MatrixFactorization model to predict the use rating or generate top item recommendations for users on the fly (Fig. 4).

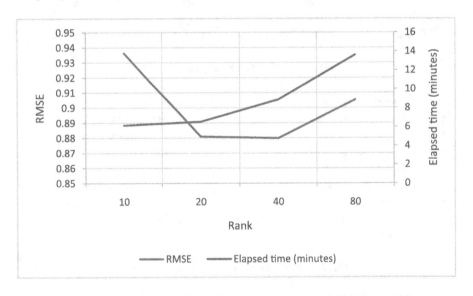

Fig. 4. Impact of increasing rank on RMSE & Performance of ALS (Lower is better)

3   Measuring Accuracy and Tuning Machine Learning Algorithms

A best practice in Machine Learning is to test the accuracy of the model on a data set that it is not trained on. In other words, we should have disjoint data sets for training and testing the models. The accuracy is reported on the test data set. We see that the data set is split for Naïve Bayes algorithm using SQL queries to segregate data but not so for Logistic Regression. We'd recommend making the measurement of accuracy consistent across all supervised ML algorithms in BigBench by either employing a method similar to Naïve Bayes to split the training/test data or using mechanisms like Cross Validation to achieve the same. This will help avoid bias and obtain a better estimate of the model's generalized error.

In a modified Logistic Regression scenario similar to Scenario #2 in Sect. 4.1 below, we split the data set into a 70% training data and 30% test data. Running prediction on the training data set vs running prediction on a 30% test data set suggests that Area Under Curve (AUC) on test data reduces by 4.4% (Fig. 5).

	Training data	Test data
Precision	0.7587515979164543	0.7281337029540884
AUC	0.5253761346454113	0.5024392467169441
Confusion Matrix	2284497.0 276466.0 478607.0 90287.0	662283.0 180646.0 74435.0 20895.0

Fig. 5. Comparison of accuracy on training data vs test data

On a related note, one common use case in Machine Learning is model selection and tuning the parameters of Machine Learning algorithms. We could modify Q05 or any new ML algorithm in BigBench to represent such a ML tuning use case. For example, test the Logistic Regression or Collaborative Filtering with multiple values of regularization parameters. The optimal regularization parameter is one that reports the best accuracy on the test data set.

4   Visualization of Machine Learning Use Case in SPSS Modeler

IBM SPSS Modeler [3] is a powerful data mining workbench that helps build accurate predictive models quickly and intuitively, without the need for any programming. It provides a rich set of machine learning algorithms and facilitates comparison of alternate ML models.

We analyzed the existing 4 Machine Language use cases in BigBench using SPSS Modeler. Our intent was to:

(a) Gain insights about the data and relative importance of features in predicting outcomes
(b) Assess the cluster quality and size for the three K-Means clustering scenarios
(c) Assess alternative models that could be incorporated into the BigBench test suite for increased coverage

We discuss our findings in the sections below.

4.1   Q05 – Logistic Regression

This use case predicts if a visitor will be interested in a given item category, based on demographics and existing users online activities (interest in items of different categories).

The target variable for the use case is a binary variable indicating whether a visitor is interested in a specific item category. It is 1 if the user's clicks in the category is greater than the average clicks in that category for the entire population.

Model Assessment. The model selected and its accuracy depend on the clicks in the specified item category. If the target category is included in the input feature vector, the model is able to predict with 100% accuracy, as it is able to learn that the clicks of the category are determining the outcome. Therefore, we suggest adding a use case in which the clicks in the item category are deliberately excluded from the input feature vector. This will enable BigBench to exercise a wider range of ML algorithms like Neural Networks and also, help increase the computational complexity of these algorithms e.g. increased depth of the Tree. The experiments below provide more details.

Scenario #1: Q05 is executed with target item category "Books". The target variable is 1 if the number of clicks by a customer is greater than average number of clicks in "Books" category and 0 otherwise. CLICKS_IN_3 column is the number of clicks in "Books" category and is a part of input feature vector.

SPSS Modeler evaluated several classification algorithms and arrived at three models: C51, Logistic Regression and C&R Tree. C51 and C&R Tree belong to the tree family. As shown below, all algorithms predict with 100% accuracy, as they are able to determine CLICKS_IN_3 as the key predictor (Fig. 6).

Graph	Model	Build Time (mins)	Max Profit	Max Profit Occurs In (%)	Lift(Top 30%)	Overall Accuracy (%)	No. Fields Used	Area Under Curve
	C5 1	15	1,564,925.0	9	3.333	100.0	1	1.0
	Logistic regression 1	15	1,564,925.0	9	3.333	100.0	9	1.0
	C&R Tree 1	15	1,564,925.0	9	3.333	100.0	6	1.0

Fig. 6. Top 3 Classification models filtered by SPSS, sorted by Area Under Curve

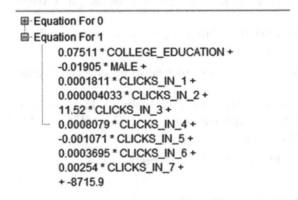

⊞ Equation For 0
⊟ Equation For 1
        0.07511 * COLLEGE_EDUCATION +
        -0.01905 * MALE +
        0.0001811 * CLICKS_IN_1 +
        0.000004033 * CLICKS_IN_2 +
        11.52 * CLICKS_IN_3 +
        0.0008079 * CLICKS_IN_4 +
        -0.001071 * CLICKS_IN_5 +
        0.0003695 * CLICKS_IN_6 +
        0.00254 * CLICKS_IN_7 +
        + -8715.9

Fig. 7. Parameters of Logistic Regression

Logistic Regression shows very high correlation for CLICKS_IN_3. Similarly, C51 is able to infer the condition that decides the target variable in a one-level tree structure (Fig. 7).

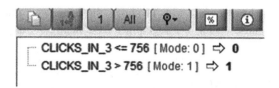

CLICKS_IN_3 <= 756 [Mode: 0] ⇨ **0**

CLICKS_IN_3 > 756 [Mode: 1] ⇨ **1**

Fig. 8. Tree output of C51 Model

Scenario #2: Q05 is executed with target item category "Toys & Games". The target variable is 1 if the number of clicks by a customer is greater than average number of clicks in "Toys & Games" category and 0 otherwise. Clicks in this category are "not" part of input feature vector (Fig. 8).

SPSS Modeler shows that three models (C51, Neural Network and Logistic Regression) have relatively high accuracy amongst several classification algorithms. The accuracy level is less than 100% (Fig. 9).

Graph	Model	Build Time	Max Profit	Max Profit Occurs in (%)	Lift(Top 30%)	Overall Accuracy (%)	No. Fields Used	Area Under Curve
✓	C5 1	4	579,106.558	8	2.896	91.156	9	0.903
✓	Neural Net 1	6	540,565	7	3.018	90.885	9	0.927
✓	Logistic regression 1	4	543,065	7	3.022	90.901	9	0.928

Fig. 9. Top 3 classification models filtered by SPSS, sorted by Area Under Curve

C51 Model. The Fig. 10 shows the decision tree structure produced by C51 Model [6]. It is worth noting that the optimal tree depth selected for producing high accuracy is 25. Adding a tree based classification algorithm to BigBench would be an interesting add-on.

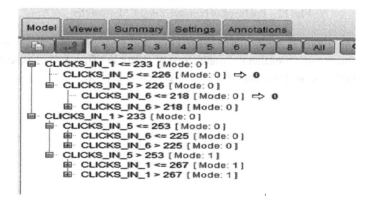

Fig. 10. Tree output by C51 Model

Logistic Regression Model. The Fig. 11 shows the parameters chosen by Logistic regression for the feature vector. Demographics have negligible impact on the target variable (Fig. 12).

```
⊞ Equation For 0
⊟ Equation For 1
        0.01647 * CLICKS_IN_1 +
        0.01627 * CLICKS_IN_2 +
        0.01658 * CLICKS_IN_3 +
        0.01667 * CLICKS_IN_4 +
        0.01651 * CLICKS_IN_5 +
        0.01646 * CLICKS_IN_6 +
        0.01629 * CLICKS_IN_7 +
        -0.003007 * [COLLEGE_EDUCATION=0] +
        0.003655 * [MALE=0] +
        + -25.8
```

Fig. 11. Parameters of Logistic Regression

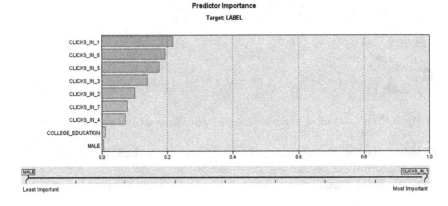

Fig. 12. Predictor importance using C51 algorithm

Predictor Importance. Both logistic regression and C51 show that the number of clicks in different categories dictates the interest of visitor in "Toys & Games" category. Demographics have the least impact on the target variable (Fig. 13).

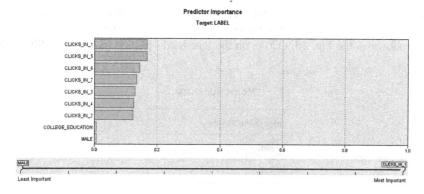

Fig. 13. Predictor importance using Logistic Regression

Similar results were observed when the target item category was "Books" and CLICKS_IN_3 was "not" included in the input feature vector (Fig. 14).

Graph	Model	Build Time (mins)	Max Profit	Max Profit Occurs in	Lift(Top 3...	Overall Accuracy	No. Fields	Area Under
	Logistic regression...	11	706,920.0	7	3.282	94.145	8	0.967
	Neural Net 1	11	699,860.0	7	3.279	94.088	8	0.966
	CHAID 1	11	641240.571	9	3.266	93.722	6	0.96

Fig. 14. Top 3 Classification models filtered by SPSS

Key Inferences.

As mentioned, not including the deterministic clicks in the input feature vector will exercise and stress the machine learning algorithms in a more realistic way. This clearly reflects in the tree depth – 25 in Scenario #2 versus only 1 in Scenario #1 Another benefit of Scenario #2 is the ability to introduce more algorithms such as Trees and Neural Networks to the BigBench ML mix.

4.2   K-Means Clustering

In SPSS, the quality of clustering is measured by the Silhouette coefficient. Silhouette coefficient combines the concepts of cluster cohesion (favoring models which contain tightly cohesive clusters) and cluster separation (favoring models which contain highly separated clusters).

Q20 performs Customer segmentation for return analysis: Customers are separated along the following dimensions: return frequency, return order ratio (total number of orders partially or fully returned versus the total number of orders), return item ratio (total number of items returned versus the number of items purchased) and return amount ratio (total monetary amount of items returned versus the amount purchased).

As shown in the Fig. 15, Q20 is on the threshold of Good (0.5) on the Silhouette scale.

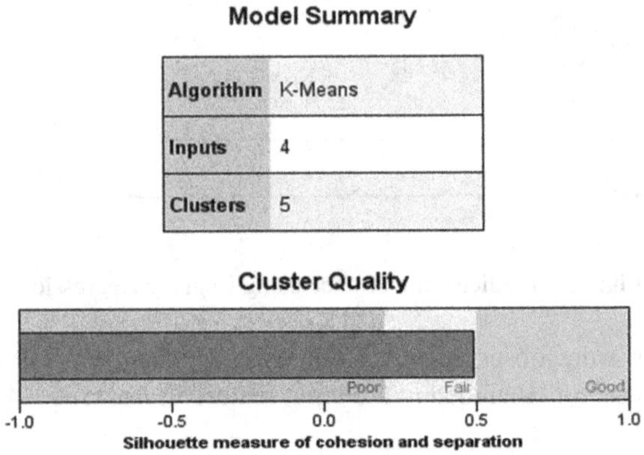

Model Summary

Algorithm	K-Means
Inputs	4
Clusters	5

Cluster Quality

Silhouette measure of cohesion and separation

Fig. 15. Cluster Quality of Q20 K-Means Clustering: twenty iterations & five Clusters

Cluster Label	cluster-5	cluster-1	cluster-4	cluster-2	cluster-3
Description					
Size	38.3% (1197745)	33.7% (1054569)	21.8% (682268)	6.1% (190173)	0.2% (5901)
Inputs	frequency 3.92	frequency 1.28	frequency 4.38	frequency 5.65	frequency 1,056.14
	itemsratio 0.06	itemsratio 0.02	itemsratio 0.11	itemsratio 0.17	itemsratio 0.05
	monetaryratio 0.03	monetaryratio 0.01	monetaryratio 0.06	monetaryratio 0.10	monetaryratio 0.03
	orderratio 0.11	orderratio 0.05	orderratio 0.17	orderratio 0.24	orderratio 0.09

Fig. 16. Q20 Clusters ordered by size and predictor importance; Cluster size = 5

Predictor Importance. In context of clustering algorithms, predictor importance indicates how well the variable can differentiate different clusters. For both range (numeric) and discrete variables, the higher the importance measure, the less likely the variation for a variable between clusters is due to chance and more likely due to some underlying difference.

In Q20, all inputs have equal importance. The size of the largest cluster is 1.2 million (38% of population) while the size of smallest cluster is 5901 (0.2% of the population). The numbers beneath the feature names are the cluster centers (Fig. 16).

Q25 performs customer segmentation based on recency of last visit, frequency of visits and monetary amount spent. Recency doesn't have any impact on the clustering, while frequency of visits and amount spent are equally important in determining the cluster.

Although Q25 fares very good on the Silhouette scale, there is a skew in cluster size with 99.7% of population belonging to a single cluster (Fig. 17).

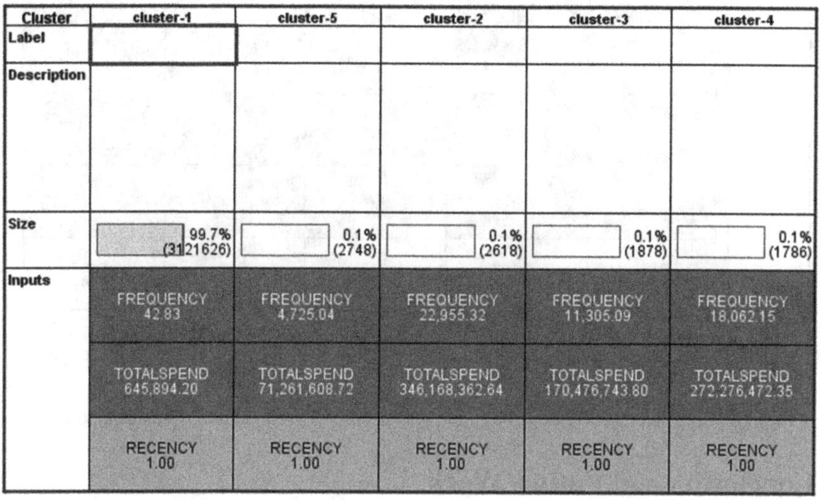

Cluster Label	cluster-1	cluster-5	cluster-2	cluster-3	cluster-4
Description					
Size	99.7% (3121626)	0.1% (2748)	0.1% (2618)	0.1% (1878)	0.1% (1786)
Inputs	FREQUENCY 42.83	FREQUENCY 4,725.04	FREQUENCY 22,955.32	FREQUENCY 11,305.09	FREQUENCY 18,062.15
	TOTALSPEND 645,894.20	TOTALSPEND 71,261,608.72	TOTALSPEND 346,168,362.64	TOTALSPEND 170,476,743.80	TOTALSPEND 272,276,472.35
	RECENCY 1.00	RECENCY 1.00	RECENCY 1.00	RECENCY 1.00	RECENCY 1.00

Fig. 17. Q25 Clusters ordered by size and predictor importance; Cluster size = 5

Q26 clusters customers into book buddies/club groups based on their store book purchasing histories. For this query, we observe non-zero store sales values for books in 5 out of 15 categories. 10 categories show 0 sales and hence they do not have any impact on the clustering.

The clustering characteristics for Q26 are similar to Q25. Although it fares very well on Silhouette scale, 99.7% of the population belongs to a single cluster (Fig. 18).

In our performance experiments, we observe that K-Means runs for the default number of iterations, twenty, for convergence. Also, K-Means requires the number of clusters to be specified. Given these factors, the skew in the cluster size should not impact the extent to which K-Means is exercised. During our experiments, we found that the Spark cache size impacts the K-Means queries significantly and hence they were effective in assessing the benefits of an optimized data connector. K-Means runs for twenty iterations and if the data doesn't fit in the memory, the analytics engine has to repeatedly fetch the input vector from the data source. The optimizations done in the dashDB Spark data connector improved performance significantly.

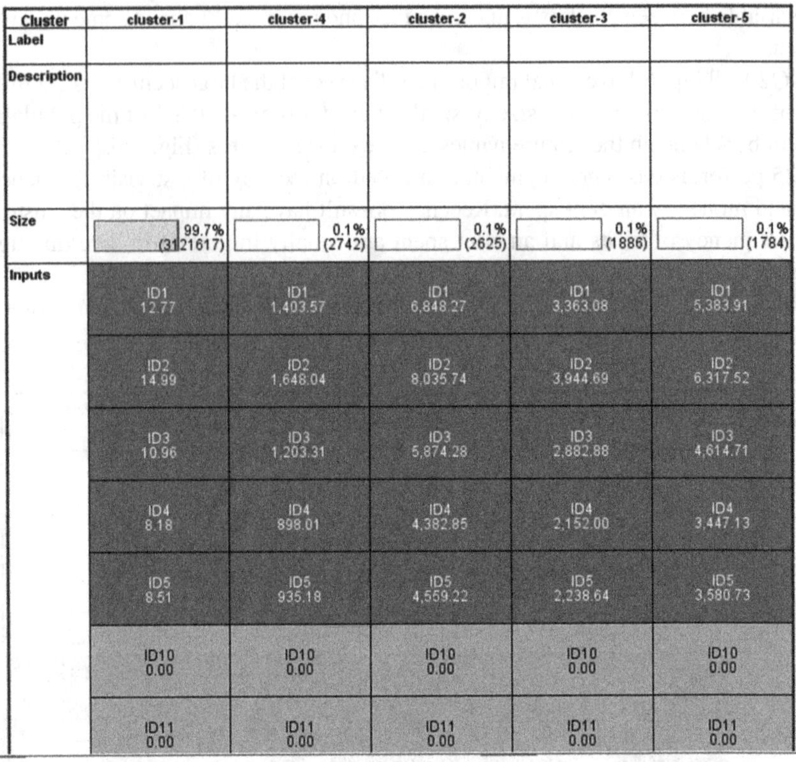

Cluster	cluster-1	cluster-4	cluster-2	cluster-3	cluster-5
Label					
Description					
Size	99.7% (3121617)	0.1% (2742)	0.1% (2625)	0.1% (1886)	0.1% (1784)
Inputs	ID1 12.77	ID1 1,403.57	ID1 6,848.27	ID1 3,363.08	ID1 5,383.91
	ID2 14.99	ID2 1,648.04	ID2 8,035.74	ID2 3,944.69	ID2 6,317.52
	ID3 10.96	ID3 1,203.31	ID3 5,874.28	ID3 2,882.88	ID3 4,614.71
	ID4 8.18	ID4 898.01	ID4 4,382.85	ID4 2,152.00	ID4 3,447.13
	ID5 8.51	ID5 935.18	ID5 4,559.22	ID5 2,238.64	ID5 3,580.73
	ID10 0.00	ID10 0.00	ID10 0.00	ID10 0.00	ID10 0.00
	ID11 0.00	ID11 0.00	ID11 0.00	ID11 0.00	ID11 0.00

Fig. 18. Q26 Clusters ordered by size and predictor importance; Cluster size = 5

5   Conclusion and Future Work

With the increasing importance of Machine Learning in big data scenarios, a broader coverage of commonly used Machine Learning algorithms along with more realistic scenarios like tuning the parameters of machine learning using cross validation and loading an existing model to predict outcome on real time data will make it even more appealing and relevant. Our experiments show that the performance characteristics of ML algorithms vary and hence will be useful in stressing the different system components. The paper highlights the importance of broadening the scope of ML algorithms and use cases in BigBench and provides concrete recommendations supported by experiments. It also talks about how the existing BigBench ML queries, K-Means and Logistics Regression, have been effective in proving the performance benefits of dashDB Spark connector.

In future, we plan to study the performance characteristics of other ML algorithms like Trees and Neural Networks on large data sets using BigBench. We'd also like BigBench to consider adopting our recommendations.

Acknowledgement. We would like to thank Berni Schiefer, Steve Rees, Torsten Steinbach, John Poelman and Manish Anand for providing their valuable feedback.

References

1. Apache Spark. http://spark.apache.org/
2. dashDB. http://www.ibm.com/analytics/us/en/technology/cloud-data-services/dashdb/
3. dashDB Local. http://www.ibm.com/analytics/us/en/technology/cloud-data-services/dashdb-local/
4. UCI Machine Learning Repository. http://archive.ics.uci.edu/ml/
5. IBM SPSS. http://www.ibm.com/analytics/us/en/technology/spss/spss.html
6. ftp://public.dhe.ibm.com/software/analytics/spss/documentation/modeler/16.0/en/modeler_applications_guide_book.pdf
7. Ghazal, A., et al.: BigBench: towards an industry standard benchmark for big data analytics. In: Proceedings of the 2013 ACM SIGMOD International Conference on Management of Data. ACM (2013)
8. Chowdhury, B., Rabl, T., Saadatpanah, P., Du, J., Jacobsen, H.-A.: A BigBench implementation in the hadoop ecosystem. In: Rabl, T., Jacobsen, H.-A., Raghunath, N., Poess, M., Bhandarkar, M., Baru, C. (eds.) WBDB 2013. LNCS, vol. 8585, pp. 3–18. Springer, Heidelberg (2014). doi:10.1007/978-3-319-10596-3_1
9. Baru, C., et al.: Discussion of BigBench: a proposed industry standard performance benchmark for big data. In: Nambiar, R., Poess, M. (eds.) TPCTC 2014. LNCS, vol. 8904, pp. 44–63. Springer, Cham (2015). doi:10.1007/978-3-319-15350-6_4
10. Nambiar, R., Poess, M. (eds.): TPCTC 2013. LNCS, vol. 8391. Springer, Heidelberg (2014). doi:10.1007/978-3-319-04936-6
11. Meng, X., et al.: Mllib: Machine learning in apache spark. JMLR **17**(34), 1–7 (2016)
12. Agrawal, D., et al.: SparkBench – a spark performance testing suite. In: Nambiar, R., Poess, M. (eds.) TPCTC 2015. LNCS, vol. 9508, pp. 26–44. Springer, Heidelberg (2016). doi:10.1007/978-3-319-31409-9_3
13. Su, X., Khoshgoftaar, T.M.: A survey of collaborative filtering techniques. Adv. Artif. Intell. **2009**, 19 (2009). Article ID 421425, doi:10.1155/2009/421425
14. Koren, Y., Bell, R., Volinsky, C.: Matrix factorization techniques for recommender systems. Computer **42**(8), 30–37 (2009)
15. Zhou, Y., Wilkinson, D., Schreiber, R., Pan, R.: Large-scale parallel collaborative filtering for the netflix prize. In: Fleischer, R., Xu, J. (eds.) AAIM 2008. LNCS, vol. 5034, pp. 337–348. Springer, Heidelberg (2008). doi:10.1007/978-3-540-68880-8_32
16. Jain, P., Netrapalli, P., Sanghavi, S.: Low-rank matrix completion using alternating minimization. In: Proceedings of the Forty-Fifth Annual ACM Symposium on Theory of Computing. ACM (2013)
17. Transaction Processing Performance Council. http://www.tpc.org
18. Zaharia, M., Chowdhury, M., Das, T., Dave, A., Ma, J., McCauley, M., Franklin, M., Shenker, S., Stoica, I.: Resilient distributed datasets: a fault-tolerant abstraction for in-memory cluster computing. In: Proceedings of the 9th USENIX Conference on Networked Systems Design and Implementation. USENIX Association, p. 2 (2012)
19. Zaharia, M., Chowdhury, M., Franklin, M.J., Shenker, S., Stoica, I.: Spark: cluster computing with working sets. In: Proceedings of the 2nd USENIX Conference on Hot Topics in Cloud Computing, Boston, 22–25 June 2010, p. 10 (2010)

20. Pilászy, I., Zibriczky, D., Tikk, D.: Fast als-based matrix factorization for explicit and implicit feedback datasets. In: Proceedings of the Fourth ACM Conference on Recommender Systems. ACM (2010)
21. Feuerverger, A., He, Y., Khatri, S.: Statistical significance of the Netflix challenge. Stat. Sci. **27**, 202–231 (2012)
22. Hastie, T., et al.: Matrix completion and low-rank SVD via fast alternating least squares. J. Mach. Learn. Res. **16**, 3367–3402 (2015)

Benchmarking Exploratory OLAP

Mahfoud Djedaini[1]([✉]), Pedro Furtado[2], Nicolas Labroche[1], Patrick Marcel[1], and Verónika Peralta[1]

[1] University of Tours, Blois, France
{mahfoud.djedaini,nicolas.labroche,patrick.marcel,
veronika.peralta}@univ-tours.fr
[2] University of Coimbra, Coimbra, Portugal
pnf@dei.uc.pt

Abstract. Supporting interactive database exploration (IDE) is a problem that attracts lots of attention these days. Exploratory OLAP (On-Line Analytical Processing) is an important use case where tools support navigation and analysis of the most interesting data, using the best possible perspectives. While many approaches were proposed (like query recommendation, reuse, steering, personalization or unexpected data recommendation), a recurrent problem is how to assess the effectiveness of an exploratory OLAP approach. In this paper we propose a benchmark framework to do so, that relies on an extensible set of user-centric metrics that relate to the main dimensions of exploratory analysis. Namely, we describe how to model and simulate user activity, how to formalize our metrics and how to build exploratory tasks to properly evaluate an IDE system under test (SUT). To the best of our knowledge, this is the first proposal of such a benchmark. Experiments are two-fold: first we evaluate the benchmark protocol and metrics based on synthetic SUTs whose behavior is well known. Second, we concentrate on two different recent SUTs from IDE literature that are evaluated and compared with our benchmark. Finally, potential extensions to produce an industry-strength benchmark are listed in the conclusion.

1 Introduction

Supporting exploration of databases is of prime importance, especially in a context of big, distributed and heterogeneous data, as shown in a recent survey of the topic [17]. Both researchers and companies that supply data analysis tools are increasingly focused on mechanisms for improving user experience, in particular aids for effective data exploration. As researchers and companies implement, test and tune alike their Interactive Data Exploration (IDE) solutions, a major issue they face is how to assess and compare solutions, improvements and alternatives.

While there exist a set of benchmarks recognized by the database community as relevant for evaluation and comparison of performance of database systems, such as the benchmarks from TPC organization, there is yet no commonly agreed upon benchmark for evaluating to what extent database systems help the user

© Springer International Publishing AG 2017
R. Nambiar and M. Poess (Eds.): TPCTC 2016, LNCS 10080, pp. 61–77, 2017.
DOI: 10.1007/978-3-319-54334-5_5

during data exploration. Our objective is to propose such a benchmark as, roughly speaking, actual TPC benchmarks assess data retrieval, and not data exploration.

In this work, we focus on the context of OLAP analysis of data, as an important use case of IDE, first because exploration of data has been deeply studied in this context, and second because to our opinion OLAP is an ideal first step before generalizing to other database systems. OLAP is defined as the process of analyzing multidimensional datasets (cubes), online, interactively, summarizing key performance indicators (called measures) from different perspectives or axes of analysis (called dimensions).

In order to motivate our work, let's now consider the following toy example: a user navigating OLAP sales cube faces an unexpected difference between sales in year 2014 and year 2015 for a product P in France. The user will then explore the surrounding region of the cube by means of OLAP operators such as roll-up (at the Europe level for example), drill-down (at the month level for example) and slices (for other products) to find evidences that may explain and corroborate the first fact. The user might even get some support from a system that automatically proposes next moves in the analysis [1,13]. We consider that the surrounding region of the first interesting fact corresponds to a neighborhood that has to be covered to ensure the exploration task success. If one wants to evaluate this particular data exploration, it is then possible to measure several metrics such as the number of queries that the user needed to cover this neighborhood, the ratio of this area that has been finally discovered, the ratio of the rest of the cube that the user had to visit to reach this result etc. So far, the assessment of data exploration through quality measures has been overlooked by the database community, but we can benefit from experience in the fields of information retrieval and exploratory search [33], which are particularly driven by the quality of the user's experience and metrics for measuring it.

This paper covers all the aspects of the implementation of a data exploration benchmark for OLAP. This benchmark was designed with a set of guiding principles in mind. It has to be easy to use by anyone, considering that a developer or researcher working in an OLAP exploration tool or algorithm, should be able to quickly plug his approach to the benchmark and use it, without requiring complex development or setting up of schema, dataset or OLAP exploration characteristics. The benchmark should also return objective evaluation metric results that are independent of the approach being tested. This means that both the mechanisms of the benchmark and the evaluation metrics must be agnostic of the IDE approach. Therefore, the benchmark includes generating skewed data with interesting facts, generating past query logs on this data and simulating OLAP users to evaluate a System Under Test (SUT) that provides support for next data analysis moves to the user. The benchmark can be used with any SUT, to evaluate any strategy that one may design. It reports objective measures for a set of metrics that characterize the degree to which the SUT fulfills certain objectives. Extensive experiments have been conducted to validate our benchmark proposal. First we evaluate the benchmark protocol and metrics based on synthetic SUTs whose behavior is well known. Second, we show that it is possible and meaningful to compare two state-of-the-art SUT from IDE literature [1,13] with our benchmark.

The paper is organized as follows. Section 2 discusses related work. Section 3 explains how interactive explorations can be scored and defines the benchmark metrics. Section 4 presents the benchmark itself. Experimental results are discussed in Sect. 5. Finally, Sect. 6 concludes the paper with considerations on potential extensions to produce industry-strength benchmark. A long version of this paper, with additional details, examples and tests, is available in [8][1].

2   Related Work

The variety of database exploration approaches. Many approaches have recently been developed to support interactive database exploration (IDE), as illustrated by a recent survey of the topic [17]. Techniques range from Visual optimization (like query result reduction [4]), automatic exploration (like query recommendation [9]), assisted query formulation (like data space segmentation [31]), data prefetching (like result diversification [19]) and query approximation [16]. The core of most of these approaches consists of a function that, given the database instance and users' history with the database (i.e., past and current queries), computes new relevant queries, tuples or visualizations that are meant to support user exploration.

Given the exploratory nature of OLAP analysis of multidimensional data (see e.g., [18,29]), many exploration techniques have been specifically developed in the context of interactive OLAP exploration of data cubes. Table 1 lists these exploration approaches, indicating their categories (in terms of those proposed in [17]), and details their inputs and outputs. For instance, the PROMISE prefetching approach [27], that predicts a query based on a Markov Model constructed out of the server's log, corresponds to a function with signature $\langle L, \langle q \rangle \rangle \rightarrow \langle q' \rangle$, where L is the query log, q is the current user query and q' is the predicted query.

Measuring the quality of an exploration. Measuring the quality of exploration has attracted a lot of attention in Information Retrieval, in particular in the field of Exploratory Search[2] [33] that can be defined as a search paradigm centered on the user and the evolution of their knowledge. It is particularly driven by the quality of the user's experience, and metrics for measuring it have been categorized as follows. **Engagement and Enjoyment** measures the "degree to which users are engaged and are experiencing positive emotions". It includes "the amount of interaction required during exploration", the "extent to which the user is focused on the task". **Task Success** assesses "whether the user reaches a particular target" and finds a "sufficient amount of information and details" along the way. **Information Novelty** measures the "amount of new

---

[1] More information on the benchmark can be found on its web page: http://www.info. univ-tours.fr/~marcel/benchmark.html.
[2] http://wp.sigmod.org/?p=1183.

Table 1. Interactive cube exploration techniques signatures

	Category	Input			Output
		DB instance	Query log	Current query	
[5]	Automatic exploration	✓		✓	Tuples
[1]	Automatic exploration		✓	✓	Sequence of queries
[12]	Automatic exploration	✓	✓	✓	Queries
[3]	Automatic exploration		✓	✓	Queries
[13]	Visual optimization	✓		✓	Queries
	Automatic exploration				Result highlighting
[14]	Visual optimization			✓	Query
[29]	Data prefetching	✓	✓	✓	Tuples
[28, 30]	Data prefetching	✓		✓	Tuples
[18]	Data prefetching	✓		✓	Sequence of queries
[27]	Data prefetching		✓	✓	Queries

information encountered". **Task Time** measures the "time spent to reach a state of task completeness". **Learning and Cognition** measures the "attainment of learning outcomes", "the amount of the topic space covered" and "the number of insights acquired". While these categories have been proposed in the context of web search, they make perfect sense for interactive database exploration, and we next focus on measures that have been proposed in the literature in these categories.

User engagement measures are popular in web search to measure how a user is engaged in using a website or search engine. Many implicit measures have been proposed [22] to track online behavior. These measures are classically categorized in activity (how a website is used), loyalty (return of users to a website) and popularity (how much a website is used). While loyalty and popularity essentially make sense for relative comparison of websites, activity enables measuring engagement for a particular website independently of other websites. The most commonly used activity metrics include number of queries per session, number of clicks, number of clicks per query, dwell (presence) time (see e.g., [10,32]).

Task success is well studied in information retrieval, with even conferences devoted specifically to this, like the TREC conference[3]. Task success is traditionally measured with precision/recall-based measures, which supposes that the target of the task is known. In this case, roughly speaking, recall measures how complete the answer to a query is, while precision measures how noisy the answer to a query is.

Many works have been interested with measuring information novelty in relational databases. For instance, in [11], the authors propose to describe the data space covered by a session with a vector of the tuples accessed by the queries of the session. In [23], the authors propose the notion of access area to capture the portion of the dataset a user is interested in. In [19], the authors use a sim-

---

[3] http://trec.nist.gov/.

ilar notion to propose query result diversification. In data cubes exploration, Sarawagi [29] assimilates novelty with the most informative constraints so that the expected distribution of a cube's cell values - based on a maximum entropy principle - is closer to the actual observed values. Here, a constraint is defined at an aggregate level of the observed cells and is expressed as a sum over the values of a subset of the observed cells. It is then expected that bringing more constraints modifies the expected distribution of values and thus allows to reduce the divergence between the observation and the expectation. The constraints that best reduce this divergence is declared to be the most informative.

Measuring task time may seem straightforward, but one needs to carefully define what is timed and how to report it. Performance related metrics like query per hour can be adapted from TPC benchmarks to this end.

Finally, measuring learning and cognition has attracted lots of attention in learner models [7]. Learner models are central components of intelligent tutoring systems, that infer a student's latent skills and knowledge from observed data. A very influential and widespread accepted model is the Knowledge Tracing model [6]. Knowledge tracing is a Bayesian network allowing to measure the probability that a skill is mastered when resolving a problem (opportunity to use the skill). The model relies on four parameters, usually experimentally tuned: $P(L_0)$: the probability the skill is already mastered before the first problem, $P(T)$: the probability the skill will be learned at each opportunity to use the skill (transition from not mastered to mastered), g: the probability the resolution is correct if the skill is not mastered (guess), s: the probability a mistake is made if the skill is mastered (slip). The probability that the skill L at opportunity n is mastered is the probability the skill is learned at step $n - 1$ or not learned at step $n - 1$ but learned at this step n. It can be computed as:

$$P(L_n|X_n = x_n) = P(L_{n-1}|X_n = x_n) + (1 - P(L_{n-1}|X_n = x_n)) \times P(T) \quad (1)$$

where:
$$P(L_{n-1}|X_n = 1) = \frac{P(L_{n-1})(1-s)}{P(L_{n-1})(1-s)+(1-P(L_{n-1}))g}$$
$$P(L_{n-1}|X_n = 0) = \frac{P(L_{n-1})s}{P(L_{n-1})s+(1-P(L_{n-1}))(1-g)}$$
$X_n = 1$ (resp. 0) means problem n has been solved (resp. not solved).

Current benchmarks for decision support, big data and analytic workloads. TPC proposes a number of popular benchmarks and metrics for assessing the performance of database systems, covering time, performance, price, availability or energy consumption (see Table 2). However, while TPC acknowledges the importance of the explorative nature of decision support queries (see e.g., the OLAP interactive queries in the TPC-DS benchmark), none of the existing TPC metrics are appropriate for measuring database exploration support in the sense of the categories proposed in Exploratory Search. A recent benchmark targets analytical workloads [21], but it too overlooks assessing the quality of interactive data exploration by proposing metrics covering only query response time, tuning overhead, data arrival to query time, storage size and monetary cost.

Table 2. Metrics of relevant TPC benchmarks

Metrics	TPC-H	TPC-DS	TPC-DI	TPCx-HS	TPCx-BB
Query per hour/minute	✓	✓		✓	✓
Price/performance	✓	✓	✓	✓	✓
Availability date	✓	✓	✓	✓	✓
Power/performance	✓	✓		✓	✓
Power	✓				
Throughput	✓	✓	✓	✓	✓
Load time		✓			✓
Power test elapsed time		✓			✓

OLAP exploration as a relevant use-case. Interestingly, the literature on OLAP already provides the building blocks for benchmarking cube exploration. OLAP has been the subject of specific benchmarks, like the TPC-H-based Star Schema Benchmark (SSB) [24]. SSB models a realistic use case of sales analysis, for which realistic instances with skewed data can be produced with the PDGF data generator [25]. Realistic OLAP workloads can be generated with the CubeLoad session generator [26]. CubeLoad takes as input a cube schema and creates the desired number of sessions according to templates modeling various user exploration patterns: users with limited OLAP skills pursuing a specific analysis goal, more advanced users navigating with a sequence of slice and drill operations, users tracking unexpected results with explorative sessions. OLAP literature also provides techniques for characterizing analytic behaviors [3,27]. In these works, the user's behavior is defined as a Markov model, whose states are built from the past queries of the user, and the transitions between states are weighted by the probability of observing a query after another in the user's query log. Finally, OLAP literature also provides characterizations of interesting data in the multidimensional space. Discovery-driven analysis of data cube [5,28–30] aimed at measuring potentially surprising data, knowing already evaluated queries. These work characterize surprising data as being groups of tuples that are connected (usually one OLAP operation apart), and that, taken altogether, appear to be meaningful (usually unexpected, in the sense of e.g., information theoretic measures).

3   Evaluating an Exploration

This section describes how interactive explorations can be scored, by implementing the metrics related to user experience identified in the previous section. We first start with presenting formally the definition of an exploration in an OLAP context. A complete formal framework, with illustrative examples, can be found in the full paper [8].

3.1   Exploration in an OLAP Context

Our benchmark incorporates the explorative and interactive nature of IDE by considering user sessions as first class citizens. We define a session $s = \langle q_1, \ldots, q_k \rangle$ of length $|s| = k$ as a sequence of k OLAP queries over a data warehouse and a log as a finite set of sessions. In what follows, a log can be associated to one particular user profile (representing this user's activity) or can represent the overall activity (being the union of all user logs).

Without loss of generality, the OLAP queries we consider are dimensional aggregate queries over a data cube [15]. A query is defined as a group by set (identifying the query granularity) and a set of Boolean predicates, one for each hierarchy. During their sessions, after each query is processed, users inspect the cube cells retrieved by the query. A cell c is an element of a cube that has a particular coordinate and a measure value. The answer to a query q, denoted $answer(q)$, is the set of retrieved cells whose coordinates are defined by the query group by set and selection predicates. Thanks to the popular OLAP operations (roll-up, drill-down, slice-and-dice), users navigate the cube by exploring cells neighborhood, querying at coarser granularity (roll-up), finer granularity (drill-down) or reaching siblings in a hierarchy. This is formalized using classical relations between cells. For two cells c and c', we note $c \succeq c'$ if c' is a roll-up of c, i.e., a coordinate of c' is an ancestor of that of c in some hierarchy, and we note $c \approx c'$ if the two cells are siblings, i.e., their coordinates differ only in one sibling position in some hierarchy.

Definition 1 (Neighborhood of a cell). *The rollup (resp., drill-down, sibling) neighborhood of a cell c is the set of all cells c' such that $c' \succeq c$ (resp. $c \succeq c'$, $c \approx c'$). The OLAP neighborhood of a cell c is the union of its rollup neighborhood, its drilldown neighborhood and its sibling neighborhood.*

The neighborhood of a group of cells C, noted $neighborhood(C)$ is the union of the neighborhoods of each cell of the group. Intuitively, the neighborhood of a group of cells defines a zone of the cube to be explored to analyze this group of cell.

A user is represented by a log, i.e., the user's past explorations. This allows to characterize a user's behavior by constructing a generative model, in the spirit of what has been successfully applied in OLAP for data prefetching [27].

Definition 2 (Generative model). *Let L be a set of sessions characterizing a user. The generative model to represent this user's behavior is a Markov model of order one, i.e., a graph $\langle S, P \rangle$ where S is the set of queries of L and $P :$ $S \times S \rightarrow [0, 1]$ denotes the probability function for the state transition, computed as $P(q_1, q_2) = \frac{sessions(\langle q_1, q_2 \rangle)}{sessions(\langle q_1 \rangle)}$ where q_1 and q_2 are queries and $sessions(s)$ gives the number of sessions where the sequence s appears.*

Definition 3 (User). *Let S be a set of sessions and x be a percentage. A user u_x is a tuple $u_x = \langle s^{log}, s^{seed}, g \rangle$ where $S = s^{log} \cup s^{seed}$, $|s^{log}| = x \times |S|$ and g is the generative model built from s^{log}.*

Finally, a task $\langle s, u \rangle$ for a session consists of a set of cells to be analyzed by a user u. This set of cells is given under the form of a session s, i.e., consists of the cells retrieved by the queries of this session. This session is based on the user's seed sessions.

3.2 Metrics

As explained in Sect. 2, the benchmark metrics follow the categorization proposed in the field of Exploratory Search [33]. For each category, we propose a primary metric and a secondary metric, with the idea that secondary metrics can be used to counterbalance primary ones. Metrics of different categories have been defined so that the overlapping between them is minimal: User engagement relates only to the number of queries, novelty to cells, task success to cell neighborhood and task time only to time. Only Learning and cognition overlaps with novelty and task success since it aims at measuring the skill of finding new and relevant information. In what follows, let $u = \langle s^{log}, s^{seed}, g \rangle$ be a user, let $t = \langle s^0, u \rangle$ be a task for user u and let $s = \langle q_1, \ldots, q_k \rangle$ be a session produced for the resolution of a task t.

User engagement and enjoyment. We use two popular and simple activity metrics used in web search: click depth as primary metric, to represent overall activity, and number of clicks per query to represent how focused this activity is. Dwell time, another popular activity metric, better fits in the Task time category. In the web search context, a click correspond to following a hyperlink (i.e., an HTTP query). In the context of the benchmark, a click corresponds to a new query. The metrics are defined as follows:

- **Query depth (QD, primary)** $= k$, i.e., the number of queries in the session, needed for resolving a task.
- **Focus (F, secondary)** $= \frac{max(\{|focus(q)| | q \in s\})}{|s|}$, where $focus(q) = \langle q, \ldots, q' \rangle \subseteq S$ such that for all $q_i, q_{i+1} \in focus(q)$ the cells retrieved by q_{i+1} are in one of the neighborhood of the cells retrieved by q_i. Intuitively, this is to measure for a query q, the length of the chain of queries starting from q that are successively distant of only one OLAP operation.

Information Novelty. Capturing user interest in the data explored can be done by measuring the access area [23]. In our context, this access area would be the set of tuples (recorded in a fact table) contributing to form the cells of a query result. As this area corresponds to tuples that are not actually presented as answers to queries (since, being an OLAP context, these tuples are aggregated), data of interest is better captured with view area, i.e., the cells presented in the answers. This is defined by: given a set of query $Q = \{q_1, \ldots, q_n\}$, the view area of Q is $va(Q) = \bigcup_{q \in Q} answer(q)$.

In a view area, not all data is interesting in the sense that it brings novel knowledge. We measure interestingness degree as a simple normalized entropy:

$interest(C) = \frac{(-\sum_{i=1}^{m} p(i) \log(p(i)))}{\log(m)}$, with $|C| = m$, $C(i)$ is the ith value of the set C and $p(i) = \frac{C(i)}{\sum_{i=1}^{m} C(i)}$ denotes the ith cell probability.

The primary metric then quantifies the amount of interesting data found in the session. The secondary metric measures the increase in view area compared to the user's log view area. They are defined as follows:

- **Relevant new information (RNI, primary)** $= 1 - avg_{q \in s}(interest(va(q)))$
- **Increase in view area (IVA, secondary)** $= \frac{|va(s) \backslash va(s^{log})|}{|va(s) \cup va(s^{log})|}$

Task Success. Intuitively, a task consists of investigating what can be said of a group of cells C coming from a task $\langle s, u \rangle$. The extent to which a task is complete consists of assessing how much of the neighborhood of this group of cells has been retrieved during the resolution of the task. A simple way of measuring it is with recall and precision. Recall is the primary metrics since consistently with exploratory search, we consider OLAP navigation as a recall oriented activity (what matters most is to minimize the number of false negative). The metrics are defined as follows, for a group of cells C:

- **Recall (R, primary)** $= \frac{|va(s) \cap neighborhood(C)|}{|neighborhood(c)|}$
- **Precision (P, secondary)** $= \frac{|va(s) \cap neighborhood(C)|}{|va(s)|}$

Task Time. Measuring task time is done by adapting metrics of existing TPC benchmarks. We need to measure the time for the SUT to produce its output and to process the queries needed for the resolution of the task. The primary metric comes from the TPC-DS benchmark and measures the number of queries per the time taken to resolve the task. The secondary metric simply measures the task elapsed time. The metrics are defined as follows:

- **Query per seconds (QpS, primary)** $= \frac{k}{\sqrt{T_o \times T_e}}$, where T_o is the overall time for the SUT to produce its outputs and T_e is the overall query execution time.
- **Task elapsed time (TET, secondary)** $= T_o + T_e$, where T_o is the overall time for the SUT to produce its outputs and T_e is the overall query execution time.

Learning and cognition. We adapt the Knowledge Tracing (KT) model to our context: we assimilate the skill mastering with the ability of finding interesting and novel information in the neighborhood N_C of a group of cells C. In other words, $X_n = 1$ if the nth query finds at least one more unknown cell of N_C where novelty increases for those cells compared to query $n - 1$. In this case, we say that the query is successful (from the learning point of view). It is 0 otherwise. The primary metric **Learning (L, primary)** is defined as in the classical KT model, see Eq. 1 in Sect. 2. The challenge is then to define the four parameters of KT: g, s, $P(L_0)$ and $P(T)$ based on the user generative model (UGM) since it represents the past of the user.

- $P(L_0)$ is the proportion of successful queries in the UGM.
- g (resp. s) is the probability in the UGM of passing from unsuccessful to successful (resp. from successful to unsuccessful) queries.
- $P(T)$ is defined as the average weighted position of successful queries in the sessions of the UGM, giving more importance to queries that happen earlier in the session.

The secondary metric measures the average progression of the learning curve. It is defined by the arithmetic mean of the proportional growth of the probabilities.

- **Learning growth rate (LGR, secondary)** =
$\frac{1}{n} \sum_{i=1}^{n} \left(1 + \frac{P(L_i|X_i=x_i) - P(L_{i-1}|X_{i-1}=x_{i-1})}{P(L_{i-1}|X_{i-1}=x_{i-1})}\right)$ where n is the session length.

4 The Benchmark

In this section we define the interface between the SUT and the benchmark and how the benchmark runs an experiment.

4.1 Interfacing with a SUT

In order to assess a SUT, the benchmark, simulates a user and interacts with the SUT. The SUT first builds its inner structures, if any, and obtains input metadata from the benchmark. Conceptually, a SUT requires as input all or part of the following parameters: the database (schema and instance), user traces (i.e., sequences of queries collected into the query log) and the active user's current exploration (a sequence of queries). Let D denote the set of all database instances for a given schema, Q denotes the set of all possible queries over this schema, S denotes the set of all sequences of queries (i.e. $Q \times Q \times \ldots \times Q$), and 2^A denotes the power-set of a set A. The functionality of a SUT can be defined generically as doing the transformation: $\langle D, 2^S, S \rangle \to S$. Once the SUT is ready, the evaluation protocol starts resolving a task, successively calling the SUT to provide suggestions.

In practice, the benchmark is a Java program where SUTs can be plugged to be evaluated. Its code and javadoc are available for SUT programmers on BitBucket[4]. Basically, a SUT is sought twice, (1) before starting the evaluation so it can initialize, and (2) whenever a next move suggestion is requested. From the benchmark point of view, SUTs are only seen as black boxes that perform what they are asked to perform, through a contract. Practically, a SUT is a class that implements an interface that exposes two functions $readMetadata$ and $nextSuggestion$. Function $readMetadata$ is called before starting the actual evaluation process, so the SUT can read and initialize its internal structure. Its parameter is a $Metadata$ object whose getters allow to access the cube, the list of users, past user traces, etc. Function $nextSuggestion$ is called many times during

---

[4] https://bitbucket.org/mdjedaini/ea-benchmark.

a task resolution. It provides to the SUT a given user and a current exploration (sequence of queries), and asks the SUT to suggest. It is the responsibility of the benchmark to orchestrate the whole process, and to make sure the functions are called with the right arguments.

4.2   How the Benchmark Works

The benchmark process is composed of three components. The first component initializes the benchmark. It generates the context: the database (i.e., the cube), some sequences of queries (i.e., the log), data skews to simulate interesting observations, and creates user profiles. You do not need to run this component if you reuse an existing context, but you can also create a new context with different schema or generation parameters.

The second component is responsible for the actual evaluation of a SUT. The evaluation is a simulation of a user's actual navigation, whereby the benchmark suggests some initial sequence of queries, asks the SUT for next move suggestions, then proposes some continuation, switches to ask the SUT again, and so on. This allows the benchmark to ask the SUT for suggestions multiple times, in multiple phases and focusing multiple view areas.

The third component is in charge of computing scores and reporting results. It considers the sessions produced with the SUT, and computes values for the quality metrics presented in Sect. 3.2.

4.3   Component 1: Benchmark Initialization

Initialization consists of the synthesis of an OLAP user environment. It consists mainly of data generation and user creation.

Data generation. An OLAP database (schema and instance) and a set of user sessions over it are firstly generated. The default database schema is the one of SSB benchmark [24], but the benchmark can be initialized with any other OLAP schema. We use CubeLoad [26] for automatically generating user sessions. Cube-Load generates realistic OLAP workloads, taking as input a cube schema and the desired number of sessions. Its templates enable the creation of a large number of sessions representing varied explorations and patterns. Finally, a realistic database instance is generated with PDGF [25]. We use the more frequent selection predicates in the log of sessions to produce data skew in the most queried zones of the cube.

Users creation. While CubeLoad enables the generation of a large workload and creates feasible exploration patterns, it does not assign sessions to specific users. We use an off-the-shelf clustering algorithm [20], using a similarity measure tailored for OLAP sessions [2] to generate "users". In this way a user is characterized by a set of sessions focusing on some zones of the cube. Each set of sessions is split in two parts: log and seed sessions. The former constitutes the user log that is exposed to the SUT, so that it can build its own knowledge for suggesting next moves. The latter, not shown to the SUT, is used to seed the benchmark tasks. The size of each user's log is ruled by a parameter. This allows the benchmark to evaluate the SUT when working with novice users versus advanced users, creating tasks with different difficulty levels, in the sense that it is more difficult for a SUT to suggest something interesting to a relatively new user. Finally, a generative model is learned from the log, inspired by techniques of the OLAP literature [3,27]. This generative model is a Markov Model that is used by Component 2, for simulating the interaction with a user.

4.4   Component 2: Evaluation of a SUT

This component is responsible for the simulation of a navigation, together with the SUT, in order to resolve a given task. A task can be seen as an exercise that has to be solved by SUTs. Tasks are created just before starting a SUT evaluation. The evaluation protocol first provides a seed session, which is a set of seed queries representing part of a navigation, as a context for continuation of the navigation. Then it asks the SUT for a first next move suggestion that consists of one or more queries. After the SUT suggestion, the benchmark decides if it accepts or refuses the suggestion (a real user would either follows the suggestion or not). The probability of discarding the suggestion is given as parameter. The following step is for the benchmark to indicate the next query (a real user may evaluate their own queries). This is done by finding the closest query in the user model to the current query, and stochastically determining the next query in the user model. This new query is then presented to the SUT to suggest again, and the process continues as such until a stop condition. The simulation is ran for a set of tasks (the number of tasks to run is a user-given parameter), and the whole process is preceded by the definition of tasks to accomplish.

4.5   Component 3: Scoring

All the queries recorded during task resolution are fed to the scoring component so it can compute a score for the SUT using the metrics defined in Sect. 3.2. For each metric, the scoring component first scores each task, and then, it aggregates scores for the SUT. In practice, a metric can be seen as a function that takes as input a task resolution (the queries that were played), and provides as output a number that represents the score of the metric for the given task.

5   Experiments

In this section we describe and report results on the experiments designed to validate the proposed benchmark. A first version of the benchmark application was coded in Java, using PDGF [25], CubeLoad [26] and Fuzzy C-medoids [20], as explained in Sect. 4. The tested SUTs were plugged to the benchmark application using the interface class. Experiments use the default schema (SSB) [24] with a scale factor of 1, a small global log of 50 sessions, 375 queries and 9 users with 50% of seed sessions. We generated 100 tasks for each SUTs to resolve. Tests were conducted on a laptop equiped with an i5-3210M CPU @ 2.50 GHz and 8 GB of RAM.

5.1   Experimental Setup

Validation. In order to test benchmark ranking, we compared three synthetic SUTs that have simple behavior, and then expected results. 'Random', the one having the worst strategy, returns purely random next move suggestions. 'Naive' generates queries that are one OLAP operation away from the previous query. It naively tries to stay close from the current query, but still chooses the next move randomly within that neighborhood. 'Cheater' uses 'insider information' in order to return good suggestions. Concretely, it generates queries containing exclusively one cell from the neighborhood N_C of cells in the seed session, which should fit the user's needs in terms of task success. The goal of this experiment is to confirm that the benchmark ranks these approaches as expected.

Benchmarking existing approaches. We created an experimental setup to compare the following approaches: CineCube [13] and Falseto [1]. CineCube is a multifaceted approach focusing on building a user-friendly sequence of explanations for the analysts. The approach highlights relevant cells in current views and explores automatically expansion into two one-distance children and two one-distance sibling queries, also summarizing the findings. Falseto is an OLAP session composition tool that implements a recommender system based on collaborative filtering. It features three phases: (i) search the log for sessions that bear some similarity with the one currently being issued by the user; (ii) extract the most relevant subsessions; and (iii) adapt the top-ranked subsession to the current user's session. As a baseline we also report the scores without a SUT, i.e. when sessions are created only by playing the user generative model ('User').

5.2   Analysis of Experimental Results

Table 3 shows the benchmark results for the tested SUTs. For each SUT, we report its average score and standard deviation for the 100 tasks, for all the benchmark primary and secondary metrics.

Validation. Regarding the three basic SUTs designed, the results globally allow us to rank 'Cheater' highest, followed by 'Naive' and 'Random' with the poorest performance, as expected. Having access to detailed insider information, 'Cheater' achieved a higher task success and it provides better learning, with a slightly higher learning curve. Theoretically, cheater should suggest all neighbor cells (recall of 1), but in practice, it is stopped by the protocol (number of chances reached). That explains that its recall is good, but not maximal. However, as it plays only queries containing the coverage of the study, increase in view

Table 3. Scores of the SUTs

	Engagement	Success			Time		Novelty		Learning	
	QD	F	R	P	QpS	TET	RNI	IVA	L	LGR
User	102	0.082	0.122	0.032	0.223	20.080	1.24E-004	0.012	0.377	0.554
stdev	0	0.053	0.229	0.082	0.264	19.706	3.38E-004	0.039	0.387	0.501
Random	102	0.030	0.189	0.002	0.0016	2260.400	3.23E-004	0.728	0.384	0.554
stdev	0	0.013	0.263	0.003	0.001	4151.932	2.33E-004	0.307	0.386	0.502
Naive	102	0.069	0.293	0.014	0.004	1464.560	7.72E-005	0.569	0.377	0.554
stdev	0	0.039	0.302	0.031	0.008	1180.270	1.10E-004	0.289	0.387	0.502
Cheater	101.4	0.029	0.538	0.119	0.014	319.600	7.83E-005	0.155	0.513	0.557
stdev	3	0.039	0.318	0.235	0.016	340.167	2.20E-004	0.277	0.489	0.504
Falseto	467	0.024	0.575	0.005	0.0005	2205.080	1.37E-004	0.737	0.376	0.559
stdev	25.855	0.001	0.344	0.003	0.0001	706.705	9.13E-005	0.236	0.386	0.506
Cinecube	184.2	0.018	0.398	0.013	0.006	2891.840	2.20E-004	0.908	0.377	0.556
stdev	51.027	0.006	0.333	0.039	0.010	4210.870	5.38E-004	0.092	0.387	0.503

area is lower. 'Random' proposes completely random jumps in the multidimensional space, which is less effective (lower task success). As it slowly contributes to task resolution, the stop condition (50 chances) stops its execution. That is why it obtains maximum query depth for all tasks (stdev = 0). Nevertheless, it randomly explores other cube zones, so consequently increases view area and increases learning at the cost of a poor precision. As expected, 'Naive' stays half-way between 'Cheater' and 'Random'. By moving always close to the current query, it was able to stay within relevant regions (so succeeding quite well). As 'Naive', it executes until the stop condition obtaining maximum query depth.

Benchmarking existing approaches. Results in Table 3 highlight the differences between Falseto and Cinecube and helps deciding which is best in which case. By definition, Falseto generates longer sessions than Cinecube as reflected by the Query Depth score. Falseto also generates queries that are not only related to the neighborhood of the last queries as Cinecube but that are based on collaborative filtering with user past sessions to recommend next analysis moves. This leads Falseto to produce more diverse queries than Cinecube. This is an advantage when it comes to explore the data as shown by the Recall of Falseto which is slightly better than that of Cinecube. However this comes at the cost of a lower precision, because it explores parts of the cube outside seed neighborhood.

When it comes to compare existing approaches with basic SUTs, we also retrieve coherent and intuitive results. Indeed, the scores allow to globally rank both Falseto and Cinecube better than Naive and worse than Cheater, while being good in some points. Indeed, contrary to Random and Naive that do not seem to effectively support data exploration, Falseto and Cinecube are clearly of great help for the user. According to User scores (i.e. user playing alone), they lead to a more complete exploration of relevant regions with more engagement and better task success.

6   Conclusion

In this paper we proposed the first benchmark for assessing OLAP exploration approaches. Modern OLAP exploration approaches are expected to suggest next moves to users, but an important question is how to evaluate the quality of such suggestions, and how to compare alternatives. Our benchmark uses state of the art techniques to generate data and user traces, and for its metrics definition. The benchmark is easy to use, requiring the SUT tester to write only a well-defined interface, and classifies the SUT according to a set of user-centric metrics. This is an important advance, since existing benchmarks focus almost exclusively on performance, cost or energy. To validate the approach, we have proved that the benchmark correctly ranks a set of strategies for which the behavior is known.

We plan to make all the details of the benchmark public for anyone to use and improve, and our long-term goal is that it serves as a building block of a more general benchmark for exploratory search over databases in general. We are currently working on turning our proposal into an industry-strength benchmark: we are detailing rules, procedures, reporting procedures and documentation; we are investigating the benchmark robustness and its sensitivity to the data and traces. We are currently studying how to use KT to aggregate our metrics to easily rank SUTs. We are also applying the benchmark to rank other existing exploratory approaches, as a way to create a regular use base.

References

1. Aligon, J., Gallinucci, E., Golfarelli, M., Marcel, P., Rizzi, S.: A collaborative filtering approach for recommending OLAP sessions. DSS **69**, 20–30 (2015)
2. Aligon, J., Golfarelli, M., Marcel, P., Rizzi, S., Turricchia, E.: Similarity measures for OLAP sessions. KAIS **39**(2), 463–489 (2014)
3. Aufaure, M.-A., Kuchmann-Beauger, N., Marcel, P., Rizzi, S., Vanrompay, Y.: Predicting your next OLAP query based on recent analytical sessions. In: Bellatreche, L., Mohania, M.K. (eds.) DaWaK 2013. LNCS, vol. 8057, pp. 134–145. Springer, Heidelberg (2013). doi:10.1007/978-3-642-40131-2_12
4. Battle, L., Stonebraker, M., Chang, R.: Dynamic reduction of query result sets for interactive visualizaton. In: International Conference on Big Data, pp. 1–8 (2013)
5. Cariou, V., Cubillé, J., Derquenne, C., Goutier, S., Guisnel, F., Klajnmic, H.: Embedded indicators to facilitate the exploration of a data cube. IJBIDM **4**(3/4), 329–349 (2009)
6. Corbett, A.T., Anderson, J.R.: Knowledge tracing: modelling the acquisition of procedural knowledge. UMUAI **4**(4), 253–278 (1995)
7. Desmarais, M.C., de Baker, R.S.J.: A review of recent advances in learner and skill modeling in intelligent learning environments. UMUAI **22**(1–2), 9–38 (2012)
8. Djedaini, M., Furtado, P., Labroche, N., Marcel, P., Peralta, V.: Assessing the effectiveness of OLAP exploration approaches. Technical report 315, June 2016. http://www.info.univ-tours.fr/~marcel/RR-DFLMP-1-062016.pdf
9. Drosou, M., Pitoura, E.: YmalDB: exploring relational databases via result-driven recommendations. VLDB J. **22**(6), 849–874 (2013)

10. Drutsa, A., Gusev, G., Serdyukov, P.: Future user engagement prediction and its application to improve the sensitivity of online experiments. In: WWW, pp. 256–266 (2015)
11. Eirinaki, M., Abraham, S., Polyzotis, N., Shaikh, N.: Querie: collaborative database exploration. TKDE **26**(7), 1778–1790 (2014)
12. Giacometti, A., Marcel, P., Negre, E., Soulet, A.: Query recommendations for OLAP discovery-driven analysis. IJDWM **7**(2), 1–25 (2011)
13. Gkesoulis, D., Vassiliadis, P., Manousis, P.: Cinecubes: aiding data workers gain insights from OLAP queries. IS **53**, 60–86 (2015)
14. Golfarelli, M., Rizzi, S., Biondi, P.: myOLAP: an approach to express and evaluate OLAP preferences. TKDE **23**(7), 1050–1064 (2011)
15. Gray, J., Chaudhuri, S., Bosworth, A., Layman, A., Reichart, D., Venkatrao, M., Pellow, F., Pirahesh, H.: Data cube: a relational aggregation operator generalizing group-by, cross-tab, and sub totals. Data Min. Knowl. Discov. **1**(1), 29–53 (1997)
16. Hellerstein, J.M., Haas, P.J., Wang, H.J.: Online aggregation. In: SIGMOD, pp. 171–182 (1997)
17. Idreos, S., Papaemmanouil, O., Chaudhuri, S.: Overview of data exploration techniques. In: SIGMOD, pp. 277–281 (2015)
18. Kamat, N., Jayachandran, P., Tunga, K., Nandi, A.: Distributed and interactive cube exploration. In: ICDE, pp. 472–483 (2014)
19. Khan, H.A., Sharaf, M.A., Albarrak, A.: Divide: efficient diversification for interactive data exploration. In: SSDBM, pp. 15:1–15:12 (2014)
20. Krishnapuram, R., Joshi, A., Nasraoui, O., Yi, L.: Low-complexity fuzzy relational clustering algorithms for web mining. IEEE-FS **9**, 595–607 (2001)
21. LeFevre, J., Sankaranarayanan, J., Hacigümüş, H., Tatemura, J., Polyzotis, N.: Towards a workload for evolutionary analytics. In: DanaC 2013, pp. 26–30 (2013)
22. Lehmann, J., Lalmas, M., Yom-Tov, E., Dupret, G.: Models of user engagement. In: Masthoff, J., Mobasher, B., Desmarais, M.C., Nkambou, R. (eds.) UMAP 2012. LNCS, vol. 7379, pp. 164–175. Springer, Heidelberg (2012). doi:10.1007/978-3-642-31454-4_14
23. Nguyen, H.V., Böhm, K., Becker, F., Goldman, B., Hinkel, G., Müller, E.: Identifying user interests within the data space - a case study with skyserver. EDBT **2015**, 641–652 (2015)
24. O'Neil, P., O'Neil, E., Chen, X., Revilak, S.: The star schema benchmark and augmented fact table indexing. In: Nambiar, R., Poess, M. (eds.) TPCTC 2009. LNCS, vol. 5895, pp. 237–252. Springer, Heidelberg (2009). doi:10.1007/978-3-642-10424-4_17
25. Rabl, T., Poess, M., Jacobsen, H., O'Neil, P.E., O'Neil, E.J.: Variations of the star schema benchmark to test the effects of data skew on query performance. In: ICPE 2013, pp. 361–372 (2013)
26. Rizzi, S., Gallinucci, E.: CubeLoad: a parametric generator of realistic OLAP workloads. In: Jarke, M., Mylopoulos, J., Quix, C., Rolland, C., Manolopoulos, Y., Mouratidis, H., Horkoff, J. (eds.) CAiSE 2014. LNCS, vol. 8484, pp. 610–624. Springer, Heidelberg (2014). doi:10.1007/978-3-319-07881-6_41
27. Sapia, C.: PROMISE: predicting query behavior to enable predictive caching strategies for OLAP systems. In: Kambayashi, Y., Mohania, M., Tjoa, A.M. (eds.) DaWaK 2000. LNCS, vol. 1874, pp. 224–233. Springer, Heidelberg (2000). doi:10.1007/3-540-44466-1_22
28. Sarawagi, S.: Explaining differences in multidimensional aggregates. In: VLDB, pp. 42–53 (1999)

29. Sarawagi, S.: User-adaptive exploration of multidimensional data. In VLDB, pp. 307–316 (2000)
30. Sathe, G., Sarawagi, S.: Intelligent rollups in multidimensional OLAP data. In: VLDB, pp. 531–540 (2001)
31. Sellam, T., Kersten, M.L.: Meet Charles, big data query advisor. In: CIDR (2013)
32. Song, Y., Shi, X., Fu, X.: Evaluating and predicting user engagement change with degraded search relevance. In: WWW, pp. 1213–1224 (2013)
33. White, R.W., Roth, R.A.: Exploratory Search: Beyond the Query-Response Paradigm. Morgan & Claypool Publishers, San Rafael (2009)

Lessons from OLTP Workload on Multi-socket HPE Integrity Superdome X System

Srinivasan Varadarajan Sahasranamam, Paul Cao$^{(\boxtimes)}$, Rajesh Tadakamadla, and Scott Norton

Hewlett Packard Enterprise, 3000 Hanover Street, Palo Alto, CA, USA
paul.cao@hpe.com

Abstract. With today's data explosion, databases have kept pace with the ever increasing demands of businesses by growing in size to accommodate peta-bytes and exa-bytes of data. This growth in data sizes is met by an equally impressive platform hardware engineering. These large enterprise systems are characterized by very large memory, I/O footprints and number of processors. These systems offer a good hardware consolidation platform, allowing traditional smaller data-bases to be consolidated on to larger and fewer x86 servers. In pursuit of efficient resource utilization, we have seen database implementations leverage technolo-gies like virtualization and containerization to improve resource utilization rates, while providing best possible isolation of workloads. Oracle database 12cR1 is an offering that enables high server resource utilization rates for database work-loads using the "Multitenant" feature. While scaling multi-tenant database work-loads from 1 to 4 sockets could be considered a modestly challenging task, scaling these workloads beyond 4 sockets (such as 8 or 16 sockets) presents new chal-lenges that have to be addressed to make the deployments more efficient. One of the main challenges to deal with on such highly NUMA (Non-Uniform Memory Access) architectures is the associated performance penalties in memory intensive workloads. Database software is primarily memory intensive, so the need for optimizing both the hardware and the software stack for best performance becomes very apparent. While many of the hardware optimizations are done via platform tunings in the BIOS (aka system firmware), an equal amount of tuning options are available to be explored and applied on the OS and the application side. In this paper, we focus primarily on the software based tunings available to users in the OS and the database. The information presented in this paper are an accumulation of learnings and observations made when trying to solve NUMA challenges during OLTP benchmarking with Oracle multitenant database deployed on a 16 socket HPE Integrity Superdome X under a Linux environment.

1 Introduction

1.1 NUMA and Its Significance in Scale-Up Architectures

As the name suggests, Non-Uniform Memory Access (NUMA) is an implementation of multiprocessor computer design that has parts of its memory local to each processor. It essentially means that different parts of memory exhibit varying latencies based on the memory topology of the system [1]. Platforms based on Intel x86 architecture can scale

© Springer International Publishing AG 2017
R. Nambiar and M. Poess (Eds.): TPCTC 2016, LNCS 10080, pp. 78–89, 2017.
DOI: 10.1007/978-3-319-54334-5_6

up to 8 sockets with Intel "glueless" architecture, and beyond 8 sockets with the help of node controllers [2]. An example of an 8 Socket "glueless" implementation would be PRIMEQUEST 2800E [3] and an example of node controller implementation would be HPE Superdome-X using XNC2 controllers along with SX3000 Fault Tolerant Crossbar (XBar) fabric [4].

In common terms, the memory that is directly attached to a processor's socket is referred to as "Socket Local Memory" or simply "Local Memory". Memory attached to an adjacent processor socket that can be reached over direct QPI links is referred to as Buddy Local memory. Please check reference [5] for more information on Intel QPI. Any memory that is attached to a processor which can be reached over a node controller or with a hop over an adjacent processor is referred to as Remote memory. Based on the architecture and physical distance, memory access latency of Remote Memory can be significantly higher than that of Local and Buddy Local memory.

Apart from memory access latencies, applications running on large NUMA systems also have to deal with cacheline contentions on a larger scale. The HPE Superdome X in our experiments with 16 Intel XeonE7-8890 v3 processors can hold data up to 720 MB in its L3 (LLC) cache (i.e. combined L3 cache size on the platform, considering all the processors, where each processor has an L3 cache of 45 MB). While there are definite benefits from the improved L3 cache size of newer processors, we also need to consider how cache line contentions could bring down performance in such large NUMA systems and see how a NUMA-aware application scheduling would help alleviate those issues.

1.2   Oracle 12c and Its "Multi-tenancy" Feature

Oracle Multitenant introduced with Oracle Database 12c is an architecture that allows a single large container database to hold multiple small pluggable databases. All these pluggable databases share the memory and background processes of their container database, enabling high density consolidation of databases compared to the previous architecture. It is critical that we understand how a container database and its constituent pluggable databases work, and align themselves with the underlying NUMA architecture for best performance results. Oracle database enables additional support for NUMA environments using the hidden parameter "_enable_NUMA_support". By default, Oracle databases are configured with NUMA support disabled. As mentioned in Oracle support document ID 864633.1, enabling or disabling NUMA support can impact application performance and caution needs to be exercised while enabling the support. Prerequisites for enabling this parameter are that underlying OS is NUMA aware and hardware is NUMA capable.

In the following sections, we have described a multitenant database deployment strategy on a 16 socket HPE Superdome X to minimize the scalability issues as described above.

This paper is organized in the following manner: Sect. 2 describes the test setup and configuration. Section 3 lists observations on the workload when run without explicit NUMA awareness. Section 4 describes the changes that were made to optimize the test setup to be more NUMA aware. Section 5 lists various tools and commands used for the tests. Section 6 summarizes the results. Section 7 provides a conclusion on the effort and Sect. 8 talks about future work.

2   Configuration

2.1   Hardware

The test setup consisted of a fully populated HPE Integrity Superdome X enclosure configured as a single 8-blade nPar or hard partition. We used Gen9 blades and each of these blades was populated with two Intel Xeon E7-8890 v3 processors and 512 GB RAM. Following table describes the complete hardware in detail:

Enclosure	Superdome X
Interconnect Modules	4 x HPE B-series 8/24c SAN witch
	2 x HPE ProCurve 6120XG 10Gbps
Hard Partition Size	8 Blades/16 CPUs/4TB RAM
CPUs per blade	2 x Intel Xeon E7-8890 v3 processors
Physical core count	288
Logical core count	576
RAM per blade	512GB
LOMs per blade	1 x FlexFabric 20Gb 2-port 630FLB
Mezzanines per blade	2 x HP QMH2672 16Gb
Storage	2 x HPE 3Par 7450 All Flash (32 TB)
LUN Raid Level	vRaid5

For storage connectivity, FC (Fiber Channel) cards were placed in mezzanine slots 2 & 3 of each blade. The enclosure was equipped with four SAN switch interconnect modules in interconnect bays 5, 6, 7, and 8. LUNs/Disks were presented to the hard partition from two all flash 3 Par 7440 arrays. As depicted in the image below, interconnect switches in bay 5 & 6 were connected to first 3 Par array and interconnect switches in bay 7 & 8 were connected to the second 3 Par array.

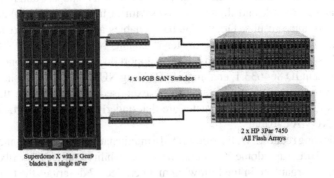

4 x 16GB SAN Switches

2 x HP 3Par 7450
All Flash Arrays

Superdome X with 8 Gen9
blades in a single nPar

2.2   OS

The Server under test was running Red Hat Enterprise Linux 6 Update 6. All the paths to a disk were grouped based on the bus and a policy for load balancing was used based on shortest service time. We had 90% of the memory configured to be used as 2 MB Huge Pages. All the necessary kernel tunables were set to support shared memory sizes required for the Oracle database. For optimal performance, we activated "latency-performance" profile using "tuned-adm" and CPU frequency governor was configured to use the "performance" profile. The test system had 16 NUMA nodes identified by the OS as indicated by "lscpu" command output. Statistics from "/proc/interrupts" indicated that interrupts for each network interface was being handled by the CPU cores from the first socket of the blade hosting the interface. For example, interrupt from network interface located on blade 2 would be handled by CPU cores from socket 2. Similarly, for a network interface card on blade 7, interrupts would be handled by CPU cores from socket 12. No other NUMA specific tunings were applied at the OS level.

2.3 Workload

We chose an open source GUI/CLI based tool called Hammer DB [6] as the workload generator for its ability to generate/reproduce results consistently. We used TPC-C like OLTP workloads and configured them to be CPU/Memory intensive to elevate the impact of remote memory access penalties and the cache line contentions. Following table describes the workload and the workload tool options used in the test environment:

HammerDB Clients	4
Workload Type	OLTP (TPC-C)
Seed Size	4000 Warehouses
Client HW Configuration	32 Cores/96 GB
Public Network	10 Gbps
User count	576
Key & Think Time	0
Workload Duration	3+5 Minutes

To be able to have comparable test results at scaled down versions, as a rule of thumb, we fixed the client side user count to the number of logical CPUs of the configuration.

2.4 Database Configuration

We had one CDB (container database) hosting four pluggable databases and each of these pluggable databases had seed data for 500 warehouses. The container database was configured to have SGA (Shared Global Area) set to consume 90% of the available memory as HugePages. After the database creation, some of the necessary parameter values were scaled to support the test environment and its workload scale. There was a single listener that was configured to support all the pluggable databases within the container database. To avoid "logfile switch" events during benchmark runs, two 2000 GB redo log groups with a single log member were configured. All the data files of the container database, datafiles of each pluggable database and redo logs were placed on different ASM diskgroups. Since this study was focused primarily on application performance impacts in a NUMA configuration, during workload runs, storage access requirements were kept to a minimum by hosting the complete database in memory while keeping the storage access limited to redo logs.

3   Observations on Non-NUMA Optimized Configuration

There are non-NUMA related factors that also significantly influence the overall performance of a database. We adopted the following sequence to present observations relevant to the subject of the paper in a definitive way:

a. Along with NUMA considerations, tune the configuration for best possible throughput. This typically involves sizing the seed data for the available memory footprint, sizing memory, tuning to resolve contentions, etc. These are standard practices for database tuning, and the Oracle specific tunings are well documented in Oracle product literature.
b. Once the baseline with the best possible results are created, switch off or rollback changes that can be classified under NUMA optimization (described in detail in Sect. 4) and make a benchmark run.

Following are a set of high-level observations that were made after step b:

1. While each HammerDB client was configured to run workload against a unique pluggable database through a single listener configured for the entire CDB, all client connections by default got clustered towards the first two NUMA nodes as indicated by PSR field of "ps–eF" output. On our 16 socket configuration, about 45% of the connections were bound to first 2 NUMA nodes while remaining were randomly distributed among the remaining 14 NUMA nodes.
2. After database startup, "numastat–p ora_" indicated near equal memory distribution across all the nodes.
3. All the db_writer (ora_db* & ora_bw*) processes were unevenly distributed(more clustered towards first few NUMA nodes).
4. Perf c2c data collected for a duration of 10 s during peak workload indicated heavy cacheline contentions with-in the database. Cache hit statistics indicated that Remote HITs + Remote HITMs contribution was around 28% while LLC misses to Remote DRAM contribution was 65%.

LLC Misses to Local DRAM	6.6%
LLC Misses to Remote DRAM	65.6%
LLC Misses to Remote cache (HIT)	4.8%
LLC Misses to Remote cache (HITM)	23.0%

5. Overall observed throughput (in TPM) was roughly equivalent to that of an 8-socket configuration running a single instance database on similar hardware. This clearly indicated that existing NUMA awareness with the configuration was insufficient to scale to 16 socket level.

4 NUMA Optimized Configuration for Best Performance

4.1 Network

As a first level of load balancing of incoming traffic, three additional listeners were configured. We configured each of these new listeners to listen on an IP plumbed on 3 different physical network interfaces. All the four network interfaces were chosen from 4 different blades of the hard partition. As an end result, we had interrupt handling for each of these four network interfaces to be handled by NUMA nodes 0, 2, 8, and 10.

4.2 Database

The first tuning in the database configuration was enabling the parameter that controlled NUMA support. Database parameter "_enable_NUMA_support" was set to true and the container database instance was restarted. With this parameter enabled, we made following two observations:

1. Distribution of db_writer processes across all NUMA nodes in a round robin fashion.
2. "buffer cache" allocation that is NUMA aligned with client side processes. Our tests strongly suggested that memory allocation for cache buffers and processes were made from the same NUMA node that handled the listeners.

4.3 Workload

Workload from each HammerDB client was targeted against a unique pluggable database using a dedicated listener configured for that pluggable database. Our second step was to isolate each of the listeners and bind them to a set of NUMA nodes to ensure that memory allocations made for the pluggable database are confined to those specific NUMA nodes.

We had LSNR1 serving the pluggable database PDB1 and was bound to NUMA nodes 0, 1, 4 & 5. Similarly, remaining listeners for the other 3 pluggable databases were bound as follows:

PDB	Listen-er	NUMA nodes
PDB 1	LSNR1	0, 1, 4, 5
PDB 2	LSNR2	2, 3, 6, 7
PDB 3	LSNR3	8, 9, 12, 13
PDB 4	LSNR4	10, 11, 14, 15

Based on HPE Superdome X architecture, either all odd slot blades or all even slot blades together form a better cluster on the XBar fabric. Hence, sharing workload between NUMA nodes 0, 1, 4, 5, 8, 9, 12, and 13 was optimal (similarly, NUMA nodes 2, 3, 6, 7, 10, 11, 14 and 15 form the other optimal cluster for sharing workload).

Tools like "numactl" and "hpe-atx" were used to bind the listeners to the assigned NUMA nodes. Unlike, "numactl", "hpe-atx" provides support for process and thread launch policies. Listeners were launched either using "numactl" or "hpe-atx" with following arguments:

```
$  numact  --cpunodebind=0,1,2,3  --membind=0,1,2,3  lsnrctl  start
LSNR1

$ hpe-atx -p ff_flat -n 0,1,4,5 lsnrctl start LSNR1
```

4.4 Summary

Idea behind each of the change was to attempt and isolate workloads against each PDB and its supporting resources to a specific set of NUMA nodes to the best extent possible.

During the course of evaluation, each of the changes described above was derived and applied to the configuration in an incremental fashion. At varying levels, we have observed substantial gains in throughput with each individual change. However, the best gains were realized with all of them applied together.

5   Tools and Commands

Following tools and commands were used to evaluate and configure the database server:

lscpu: Display information of CPU architecture along with information on CPU count, threads, cores, sockets, and Non-Uniform Memory Access (NUMA) nodes.

lstopo: Command to display the hardware topology of the system.

hwloc-distances: Displays distance matrices attached to the system topology.

numastat: Show per-NUMA-node memory statistics for processes and the operating system. The default numastat statistics shows per-node numbers (in units of pages of memory) in the categories numa_hit, numa_miss, numa_foreign, interleave_hit, local_node, other_node.

numactl: Control NUMA policy for processes or shared memory. It runs processes with a specific NUMA scheduling or memory placement policy. The policy is set for command and inherited by all of its children.

hpe-atx [7]: Similar to numactl, this is a utility that allows NUMA unaware applications to gain the benefits of NUMA, by launching processes in a more controlled manner to optimize its memory allocation and scheduling. It controls the distribution of an application's processes and threads in a NUMA environment. This tool is available on RHEL and SLES distributions supported on HPE Integrity Superdome X.

perf, c2c [8]: Enhancement added to perf tool to analyze cache line contention on NUMA systems.

6   Results

With the changes described in Sect. 4 and using "numactl", we observed a gain of 43.4% in throughput. And with "hpe-atx", a gain of 58.3%.

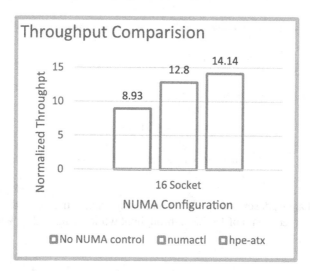

Perf c2c data for the tuned configuration indicated significant reduction in cacheline contentions and LLC misses to remote DRAM. Following are the cache level statistics for the tuned configuration. We have to note that workload against each PDB was still confined to 4 NUMA nodes where local memory to remote memory still stands at 1:3 ratio i.e. Buddy Local/Remote Memory contributes to 75% of the memory used by a PDB.

LLC Misses to Local DRAM	21.2%
LLC Misses to Remote DRAM	64.2%
LLC Misses to Remote cache (HIT)	2.8%
LLC Misses to Remote cache (HITM)	11.9%

We clearly see a 14.6% rise in LLC missed to local DRAM and a drop of 11.1% in remote cache HITMs. Apart from a rise in LLC hits and Local HITMs, we also have to note that all the LLC misses to remote DRAM, remote HITs and remote HITMs are now confined to a set of 3 NUMA nodes that are closely co-located on the Xbar fabric with ability to resolve them at much lower latency rates. From these statistics and the observed throughput improvements, the benefits of localizing memory accesses to NUMA nodes that are physically closer (with reduced performance penalties of remote cache accesses) become obvious.

When the same configuration was scaled down to 8 sockets with proportionately scaled down PDB and user counts, we achieved a scalability factor of 1.77× between 8 socket and 16 socket configurations in conjunction with "hpe-atx" tool for NUMA controlled launch of listeners:

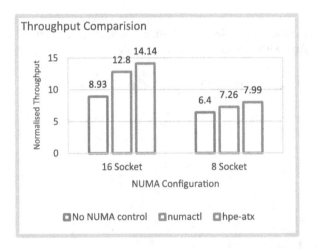

With the changes described in Sect. 4 and using "numactl", on an 8 socket configuration, we observed a gain of 13.5% in throughput while a gain of 24.8% was achieved using "hpe-atx".

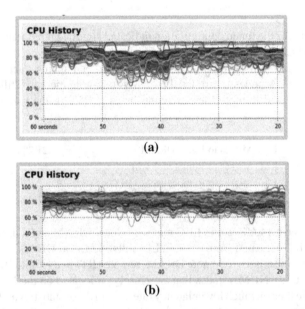

Fig. 1. a. CPU utilization pattern on 16 socket multitenant database when 4 listeners were launched with no NUMAcontrol. **b.** CPU utilization pattern on 16 socket multitenant database when 4 listeners were launched using hpe-atx.

When the listeners are launched without explicit NUMA control, utilization pattern across the CPU cores was observed to be erratic. Using following line histograms (Fig. 1a and b) from benchmark runs, we can visualize CPU utilization pattern between default and NUMA controlled launch of database listeners on our 16 socket server (with 576 logical CPUs):

With even distribution of workload across all NUMA nodes (Fig. 1b), we see steady CPU utilization rates in the 70%–90% band. And in case of no NUMA control (Fig. 1a), we see some CPUs close to 100% utilization rates while some CPUs between 70%–95% band with a lot of fluctuation.

These observations can be better visualized from following CPU utilization line histograms (Fig. 2a and b) captured from benchmark runs on an 8 socket, 120 core single instance database using a single listener. This configuration is outside the focus of this paper, but it has still been included here as it provides better insights into CPU utilization randomness between default launch and the NUMA controlled launch of listeners.

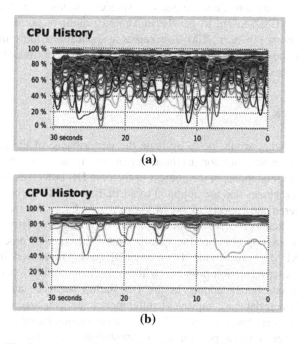

(a)

(b)

Fig. 2. a. CPU utilization pattern on 8 socket single instance database when listener was launched with no NUMA control. **b.** CPU utilization pattern when listener were launched using "hpe-atx" distributing workloads across all NUMA nodes

We can clearly observe the improvement of individual CPU core utilization levels due to even distribution of workload.

7   Conclusion

In this paper, we saw how the present day NUMA architectures affect application perform-ance. We have seen how a NUMA-unaware application, left to the mercy of the default scheduling policy and memory allocation of the OS, might suffer severe performance penalties due to remote memory accesses. We have also seen how the OS, associated tools and the software eco-system have evolved over time to be more aligned with the hardware topology of the platform and hence offer ways to overcome those performance penalties with the use of NUMA-aware scheduling and allocation policies (tools like numactl and hpe-atx serve as good examples in this area). We also discussed how these performance penalties become aggravated on large scale-up architectures, and how a NUMA-aware OS and its associated tools become even more important and relevant for applications in such environments. Our study also suggests that even though OS's are NUMA-aware, at best they can enable applications to exploit the benefits of data locality. While the OS-based NUMA-awareness may provide some gains on smaller systems, it requires applications also to be fully NUMA-aware to get the maximum benefit of performance on large systems. In the absence of such NUMA-awareness in applications, tools like numactl and hpe-atx build on the OS capabilities and play a vital role as performance enablers in an increasingly complex hardware and software eco-system.

8   Futurework

A future area of work would be to evaluate the performance of the database in the presence of storage accesses for database operations. In addition, an assessment of impacts of interrupt latencies in the context of the core DB processes (the shared pool of background processes that are common to all PDBs) would also be recommended. This study covered the aspects of NUMA-aware scheduling for the DB processes launched by the listener processes. While this has its demonstrated performance benefits, we believe there are opportunities for optimization if the core DB processes could be launched with NUMA optimization. The present DB startup framework in Oracle limits the amount of experimentation possible (without resorting to developer hacks, which requires an intimate understanding of the DB startup sequence and its associated processes). As Oracle database improves to offer more scaling on large systems, it would be worthwhile to keep this as a potential area of investigation based on future features that might lend themselves to performance optimization in this area.

Acknowledgments. We would like to thank Long, Wai Man for his help in setup of perf, c2c tool along with analysis of perf data collected during the database workload runs.

References

1. Manchanda, N., Anand, K.: Non-uniform memory access (NUMA). New York University
2. Kumar, M., Demshki, M., Shiveley, R.: Advanced reliability for intel xeon processor-based servers, March 2010

3. FUJITSU Server PRIMEQUEST 2800E Mission Critical data sheet: https://sp.ts.fujitsu.com/dmsp/Publications/public/ds-pq-2800e.pdf
4. HPE Integrity Superdome X system architecture and RAS a technical white paper
5. An Introduction to the Intel® QuickPath Interconnect, January 2009
6. HammerDB is a graphical open source database load testing and benchmarking tool for Linux and Windows. http://www.hammerdb.com/about.html
7. HPE Application Tuner Express. https://h20392.www2.hpe.com/portal/swdepot/displayProductInfo.do?productNumber=HPE-ATX
8. New tool to analyze cacheline contention on NUMA systems. https://lwn.net/Articles/588866/

Benchmarking Distributed Stream Processing Platforms for IoT Applications

Anshu Shukla$^{(\boxtimes)}$ and Yogesh Simmhan

Indian Institute of Science, Bangalore, India
shukla@grads.cds.iisc.ac.in, simmhan@cds.iisc.ac.in

Abstract. Internet of Things (IoT) is a technology paradigm where millions of sensors monitor, and help inform or manage, physical, environmental and human systems in real-time. The inherent closed-loop responsiveness and decision making of IoT applications makes them ideal candidates for using low latency and scalable stream processing platforms. Distributed Stream Processing Systems (DSPS) are becoming essential components of any IoT stack, but the efficacy and performance of contemporary DSPS have not been rigorously studied for IoT data streams and applications. Here, we develop a benchmark suite and performance metrics to evaluate DSPS for streaming IoT applications. The benchmark includes 13 common IoT tasks classified across functional categories and forming micro-benchmarks, and two IoT applications for statistical summarization and predictive analytics that leverage various dataflow patterns of DSPS. These are coupled with stream workloads from real IoT observations on smart cities. We validate the benchmark for the popular Apache Storm DSPS, and present the results.

Keywords: Stream processing · Benchmark · Workload · Internet of Things · Smart cities · Fast data · Big Data · Velocity · Distributed systems

1 Introduction

Internet of Things (IoT) is a technology paradigm where ubiquitous sensors numbering in the billions will be able to monitor physical infrastructure, humans and virtual entities in real-time, process both real-time and historic observations, and take actions that improve the efficiency and reliability of systems, or the comfort and lifestyle of society. Besides affordable sensing and pervasive communications, Cloud and Big Data platforms have contributed to this rapid growth.

Currently, the IoT applications are often manifest in vertical domains, such as demand-response optimization and outage management in *smart grids* [5], or fitness and sleep tracking by *smart watches and health bands* [19]. The IoT stack for such domains is tightly integrated to serve specific needs, but typically operates on a closed-loop *Observe Orient Decide Act (OODA)* cycle, where sensors communicate time-series observations of the system to a Cloud data center for analysis, and the analytics drive recommendations that are enacted on, or

© Springer International Publishing AG 2017
R. Nambiar and M. Poess (Eds.): TPCTC 2016, LNCS 10080, pp. 90–106, 2017.
DOI: 10.1007/978-3-319-54334-5_7

notified to, the system to improve it, which is again observed and so on. In fact, this *closed-loop* responsiveness is an essential characteristic of IoT applications.

This low-latency cycle makes it necessary to process data streaming from sensors at fine spatial and temporal scales, in *real-time*, to derive actionable intelligence. In particular, this streaming analytics has to be done at massive scales (millions of sensors, thousands of events per second) from across distributed sensors, requiring large computational resources. *Cloud computing* offers a natural platform for scalable processing of the observations at globally distributed data centers, and sending a feedback response to the IoT system at the edge. This complements *Fog Computing* that puts the onus on edge devices to collaboratively collect, process and analyze data with low latency by reduced reliability.

Recent *Big Data platforms* like Storm [18], Flink [2] and Spark [20] provide an intuitive programming model for composing and executing scalable streaming applications on commodity clusters and Clouds. These *Distributed Stream Processing Systems (DSPS)* are becoming essential components of any IoT stack to support online analytics for IoT applications. In fact, reference IoT solutions from Cloud providers[1,2] include their own stream and event processing engines.

Shared-memory stream processing systems [9] have been investigated for wireless sensor networks, with community benchmarks being proposed [6]. But there has not been a detailed review of, or benchmarks for, *distributed* stream processing for IoT domains. In particular, the efficacy of contemporary DSPS, originally designed for web and social network traffic [18], have not been rigorously studied for *IoT data streams and applications*. We address this gap here.

We develop a benchmark suite for DSPS to evaluate their effectiveness for streaming IoT applications. The proposed workload is based on common building-block tasks observed in various IoT domains for real-time decision making, and the input streams are sourced from real IoT observations from smart cities.

Specifically, we make the following contributions:

1. We classify different characteristics of streaming applications and their data sources, in Sect. 3. We propose categories of tasks that are essential for IoT applications and the key features that are present in their input data streams.
2. We identify performance metrics of DSPS that are necessary to meet the latency and scalability needs of streaming IoT applications, in Sect. 4.
3. We propose an IoT Benchmark for DSPS based on representative *micro-benchmark tasks*, drawn from the above categories, in Sect. 5. Further, we design two reference IoT applications – for *statistical analytics* and *predictive analytics* – composed from these tasks. We also offer real-world streams with different distributions on which to evaluate them.
4. We run the benchmark for the popular Apache Storm DSPS, and present empirical results for the same in Sect. 6.

---

[1] https://aws.amazon.com/iot/how-it-works/.

[2] https://www.microsoft.com/en-in/server-cloud/internet-of-things/overview.aspx.

2 Background and Related Work

Early data stream management systems (DSMS) were motivated by sensor network applications, that have similarities to IoT [9]. They supported continuous query languages with operators such as join, aggregators similar to SQL, but with a temporal dimension using windowed-join operations. These have distributed implementations [8] and have evolved to complex event processing (CEP).

Current DSPS like Apache Storm and Apache Spark Streaming [18,20] leverage Big Data fundamentals, running on commodity clusters and Clouds, offering weak scaling, ensuring robustness, and supporting fast data processing over thousands of events per second. They do not support native query operators and instead allow users to plug in their own logic composed as dataflow graphs executed across a cluster. While developed for web and social network applications, such fast data platforms have found use in financial markets, astronomy, and particle physics. IoT is one of the more recent domains to consider them.

Work on DSMS spawned the Linear Road Benchmark (LRB) [6] that was proposed as an application benchmark. In the scenario, DSMS had to evaluate toll and traffic queries over event streams from a virtual traffic monitoring system, with parallels to current smart transportation. However, there have been few studies or community efforts on benchmarking DSPS, other than research evaluations against popular DSPS like Storm or Spark. These papers define their own metrics of success – typically just throughput and latency – and use generic workloads like the Enron email dataset and custom micro-benchmarks [15].

Stream Bench [14] has proposed 7 micro-benchmarks on 4 different synthetic workload suites generated from real-time web logs and network traffic to evaluate DSPS. Metrics including performance, durability and fault tolerance are proposed. It covers different dataflow patterns and common tasks like grep and wordcount. While useful as a generic streaming benchmark, it does not consider aspects unique to IoT applications and streams. SparkBench [3] is specific to Spark, and includes four categories of applications from domains spanning Graph analysis and SQL queries, and one application for Spark Streaming. The benchmark metrics include CPU, memory, disk and network IO, with the goal of identifying tuning parameters to improve Spark's performance.

In contrast, the goal for this paper is to develop relevant micro- and application-level benchmarks for evaluating DSPS, specifically for *IoT workloads* for which such platforms are increasingly being used. Our benchmark is designed to be *platform-agnostic, simple* to implement and execute within diverse DSPS, and *representative* of both the application logic and data streams observed in IoT domains. This allows for the performance of DSPS to be independently and reproducibly verified for IoT applications.

There has been a slew of Big Data benchmarks for processing high volume (i.e., MapReduce-style) and enterprise/web data, that complement our work. The *Yahoo Cloud Serving Benchmark (YCSB)* [10] was developed to compare different key-value stores on the Cloud. *Hibench* [13] is a workload suite for evaluating Hadoop with popular micro-benchmarks like Sort, WordCount and Tera-Sort, MapReduce applications like Nutch Indexing and PageRank, and machine learning algorithms like K-means Clustering. This is a general purpose workload for MapReduce platforms at large. *BigBench* [12] uses a synthetic generator to simulate online retail enterprise data. It combines structured data from the TPC-DS benchmark [16], semi-structured data on user clicks, and unstructured data from product reviews. Queries cover data *velocity* by processing periodic data refreshes, *variety* by including free-text reviews, and *volume* by querying over a large click logs. We take a similar approach to benchmark fast data platforms, targeting the IoT domain and using real public data streams.

There has been some recent work on benchmarking IoT applications. Generating large volumes of synthetic sensor data with realistic values is challenging, yet required for benchmarking. *IoTAbench* [7] provides a scalable synthetic generator of time-series datasets using a Markov chain model for scaling the time series. It uses a limited number of inputs to ensure that important statistical properties of the stream is retained in the generated data. This has been demonstrated for smart meter data. Their emphasis is on the data characteristics and content, which supplements our focus on the systems aspects of the platform.

CityBench [4] is a benchmark to evaluate RDF stream processing systems. They include different generation patterns for smart city data, such as traffic vehicles, parking, weather, pollution, cultural and library events, with changing event rates and playback speeds. They propose fixed set of semantic queries over this dataset, with concurrent execution of queries and sensor streams. Here, the target platform is different (RDF database), but in a spirit as our work.

3   Characteristics of Streaming IoT Applications

In this section, we review the common application composition capabilities of DSPS, and the dimensions of the streaming applications that affect their performance on DSPS. These semantics help define and describe streaming IoT applications based on DSPS capabilities. Subsequently in this section, we also categorize IoT tasks, applications and data streams based on the domain requirements. Together, these offer a search space for defining workloads that meaningfully and comprehensively validate IoT applications on DSPS.

3.1   Dataflow Composition Semantics

DSPS applications are commonly composed as a *dataflow graph*, where vertices are user provided *tasks* and directed edges are refer to *streams of messages* that can pass between them. *Messages* (or events or tuples) from/to the stream are consumed/produced by the tasks. DSPS typically treat the messages as opaque content, and only the user logic may interpret the message content.

Selectivity ratio, also called *gain*, is the number of output messages emitted by a task on consuming a unit input message, expressed as $\sigma = input\ rate : output\ rate$. Based on this, one can assess whether a task amplifies or attenuates the incoming message rate. It is important to consider this while designing benchmarks as it can have a multiplicative impact on downstream tasks.

Fig. 1. Common task patterns and semantics in streaming applications.

There are message generation, consumption and routing semantics associated with tasks and their dataflow composition. Figure 1 captures the basic *composition patterns* supported by modern DSPS. Source tasks have only outgoing edge(s), and these tasks encapsulate user logic to generate or receive the input messages that are passed to the dataflow. Likewise, Sink tasks have only incoming edge(s) and these tasks react to the output messages from the application, say, by storing it or sending an external notification.

Transform tasks, sometimes called *Map* tasks, generate one output message for every input message received ($\sigma = 1 : 1$). Their user logic performs a transformation on the message, such as changing the units or projecting only a subset of attribute values. Filter tasks allow only a subset of messages that they receive to pass through, optionally performing a transformation on them ($\sigma = N : M$, $N \geq M$). Conversely, a FlatMap consumes one message and emits multiple messages ($\sigma = 1 : N$). An Aggregate pattern consumes a *window* of messages, with the window width provided as a *count* or a *time* duration, and generates one or more messages that is an aggregation over each message window ($\sigma = N : 1$).

When a task has multiple outgoing edges, routing semantics on the dataflow control if an output message is *duplicated* onto all the edges, or just one downstream task is selected for delivery, either based on a *round robin* behavior or using a *hash function* on an attribute in the outgoing message to decide the target task. Similarly, multiple incoming streams arriving at a task may be *merged* into a single interleaved message stream for the task. Or alternatively, the messages coming on each incoming stream may be conjugated, based on order of arrival or an attribute exposed in each message, to form a *joined* stream of messages.

Tasks may be *data parallel*, in which case, it may be allocated multiple threads/cores to process messages in parallel by different instances the task. This is typically possible for tasks that do not maintain state across multiple messages. The *length of the dataflow* is the latency of the critical (i.e., longest) path through the dataflow graph, if the graph does not have cycles. This gives an estimate of the expected latency for each message and also influences the number of network hops a message on the critical path has to take in the cluster.

3.2 Input Data Stream Characteristics

We list a few characteristics of the input data streams that impact the runtime performance of streaming applications, and help classify IoT message streams.

The *input throughput* in messages/sec is the cumulative frequency at which messages enter the source tasks of the dataflow. Input throughputs can vary by application domain, and are determined both by the number of streams of messages and their individual rates. This combined with the dataflow selectivity will impact the load on the dataflow and the output throughput.

Throughput distribution captures the variation of input throughput over time. In real-world settings, the input data rate is usually not constant and DSPS need to adapt to this. There may be several common data rate distributions besides a *uniform* one. There may be *bursts* of data coming from a single sensor, or a coordinated set of sensors. A *saw-tooth* behavior may be seen in the ramp-up/-down before/after specific events. *Normal* distribution are seen with diurnal (day vs. night) stream sources, with *bi-modal* variations capturing peaks during the morning and evening periods of human activity.

3.3   Categories of IoT Tasks and Applications

Here, we attempt to categorize common IoT processing and analytics tasks that are performed over real-time data streams to support domain applications.

Parse. Messages are encoded on the wire in a standard text-based or binary representation by the stream sources, and need to be parsed upon arrival at the application. Text formats in particular require string parsing by the tasks, and are also larger in size on the wire. The tasks within the application may themselves retain the incoming format in their streams, or switch to another format or data model, say, by projecting a subset of the fields. Industry-standard formats that are popular for IoT domains include CSV, XML and JSON text formats, EXI and CBOR binary formats, and serialization protocols like Google's Protocol Buffer and Apache Thrift.

Filter. Messages may require to be filtered based on specific attribute values present in them, as part of data quality checks, to route a subset of message types to a part of the dataflow graph, or as part of their application logic. Value and band-pass filters that test an attribute's *numerical value ranges* are common, and are both compact to model and fast to execute. Since IoT event rates may be high, more efficient Bloom filters may also be used to process *discrete values* with low space complexity but with a small fraction of false positives.

Statistical Analytics. Groups of messages within a sequential time or count window of a stream may require to be aggregated as part of the application. The aggregation function may be *common mathematical operations* like average, count, minimum and maximum. They may also be *higher order statistics* such as finding outliers, quartiles, second and third order moments, and counts of distinct elements. Statistical *data cleaning* like linear interpolation or denoising using Kalman filters are common for sensor-based data streams. Some tasks may maintain just local state for the window width (e.g., local average) while others may maintain state across windows (e.g., moving average). When the state size grows, here again approximate aggregation algorithms may be used.

Predictive Analytics. Predicting future behavior of the system based on past and current messages is an important part of IoT applications. Various statistical and machine-learning algorithms may be employed for predictive analytics over sensor streams. The *predictions* may either use a recent window of messages to estimate the future values over a time or count horizon in future, or train models over streaming messages that are periodically used for predictions over the incoming messages. The *training* itself can be an online task that is part of an application. For e.g., linear regression use statistics to predict uni- or multivariate attribute values. Classification algorithms like decision trees and neural networks can be trained to map discrete values to a category, which may lead to specific actions taken on the system.

Pattern Detection. Another class of tasks are those that identify patterns of behavior over several events. Unlike window aggregation which operate over static window sizes and perform a function over the values, pattern detection matches user-defined predicates on messages that may not be sequential or even span streams, and returned the matched messages. These are often modeled as *state transition automata* or *query graphs*. Common patterns include contiguous or non-contiguous sequence of messages with specific property on each message (e.g., high-low-high pattern over 3 messages), or a join over two streams based on a common attribute value. Complex Event Processing (CEP) engines [17] may be embedded within the DSPS task to match these patterns.

Visual Analytics. Other than automated decision making, IoT applications often generate *charts and animations* for consumption by end-users or system managers. These visual analytics may be performed either at the client, in which case the processed data stream is aggregated and provided to the users. Alternatively, the streaming application may itself periodically generate such plots and visualizations as part of the dataflow, to be hosted on the web or pushed to the client. Charting libraries like D3.js or JFreeChart may be used for this.

IO Operations. Lastly, the IoT dataflow may need to access external storage or messaging services to access/push data into/out of the application. These may be to store or load trained models, archive incoming data streams, access historic data for aggregation and comparison, and subscribe to message streams or publish actions back to the system. These require access to *file storage, SQL and NoSQL databases, and publish-subscribe messaging systems.* Often, these may be hosted as part of the Cloud platforms themselves.

The tasks from the above categories, along with other domain-specific tasks, are composed together to form streaming IoT dataflows. These domain dataflows themselves fall into specific classes based on common use-case scenarios, and loosely map to the Observe-Orient-Decide-Act (OODA) phases.

Extract-Transform-Load (ETL) and Archival applications are front-line "observation" dataflows that receive and pre-process the data streams, and if necessary, archive a copy of the data offline. Pre-processing may perform data format transformations, normalize the units of observations, data quality checks to remove invalid data, interpolate missing data items, and temporally reorder messages arriving from different streams. The pre-processed data may be archived to table storage, and passed onto subsequent dataflow for further analysis.

Summarization and Visualization applications perform statistical aggregation and analytics over the data streams to understand the behavior of the IoT system at a coarser granularity. Such summarization can give the high-level pulse of the system, and help "orient" the decision making to the current situation. These tasks are often succeeded by visualizations tasks in the dataflow to present it to end-users and decision makers.

Prediction and Pattern Detection applications help determine the future state of the IoT system and "decide" if any reaction is required. They identify patterns of interest that may indicate the need for a correction, or trends based on current behavior that require preemptive actions. For e.g., an unsustainable growing load on a power grid cause load to be shed preemptively, or a detection that the heart-rate from a fitness watch is very high may trigger a treadmill to slow down.

Classification and notification applications determine specific "actions" that are required and communicate them to the IoT system. Decisions may be mapped to specific actions, and the entities in the IoT system that can enact those be notified. For e.g., the need for load shedding in the power grid may map to specific residents to request the curtailment from, or the need to reduce physical activities may lead to a treadmill being notified to reduce the speed.

3.4 IoT Data Stream Characteristics

IoT data streams are often generated by sensors that observe physical systems or the environment. As a result, they are typically time-series data that are generated periodically. The sampling rate for these sensors may vary from once a day to hundreds per second, depending on the domain. The number of sensors themselves may vary from a few hundred to millions as well. As a result, we may encounter a wide range of input throughputs from 10^{-2} to 10^5 messages/sec.

At the same time, this event rate itself may not be uniform across time. Sensors may also be configured to emit data only when there is a change in observed value, rather than unnecessarily transmitting data that has not changed. This helps conserve network bandwidth and power for constrained devices when the observations are slow changing. Further, if data freshness is not critical to the application, they may sample at high rate but transmit at low rates but in a burst mode. Example smart meters may collecting kWh data at 15 min intervals from millions of residents but report it to the utility only a few times a day, while the FitBit smart watch syncs with the Cloud every few minutes or hours even as data is recorded every few seconds.

Message variability also comes into play when human-related activity is being tracked. Diurnal or bimodal event rates are seen with single peaks in the afternoons, or dual peaks in the morning and evening. For e.g., sensors at businesses may match the former while traffic flow sensors may match the latter.

4 Performance Metrics

We identify and formalize commonly-used quantitative performance measures for evaluating DSPS for the IoT workloads.

Latency. Latency for a message that is generated by task is the time in seconds it took for that task to process one or more inputs to generate that message. When we consider the *average latency* $\overline{\lambda}$ of the dataflow application, it is the average of the time difference between each message consumed at the source tasks and all its causally dependent messages generated at the sink tasks.

The latency per message may vary depending on the input rate, resources allocated to the task, and the type of message being processed. While this task latency is the inverse of the mean throughput, the *end-to-end latency* for the task within a dataflow will also include the network and queuing time to receive a tuple and transmit it downstream.

Throughput. The output throughput is the aggregated rate of output messages emitted out of the sink tasks, measured in messages per second. The throughput of a dataflow depends on the input throughput and the selectivity of the dataflow, provided the resource allocation and performance of the DSPS are adequate. Ideally, the output throughput $\omega^o = \sigma \times \omega^i$, where ω^i is the input throughput for a dataflow with selectivity σ. It is also useful to measure the *peak throughput* that can be supported by a given application, which is the maximum stable rate that can be processed using a fixed quanta of resources.

Both throughput and latency measurements are relevant only under *stable conditions* when the DSPS can sustain a given input rate.

Jitter. The ideal output throughput may deviate due to variable rate of the input streams, change in the paths taken by the input stream through the dataflow (e.g., at a `Hash` pattern), or performance variability of the DSPS. We use jitter to track the variation between the expected and observed output throughput, defined for a time interval t as, $J_t = \frac{\omega^o - \sigma \times \omega^i}{\sigma \times \overline{\omega^i}}$, where the numerator is the observed difference between the expected and actual output rate during interval t, and the denominator is the expected long term average output rate given a long-term average input rate $\overline{\omega^i}$. In an ideal case, jitter will tend towards zero.

CPU and Memory Utilization. Streaming IoT dataflows are expected to be resource intensive, and the ability of the DSPS to use distributed resources with minimal overhead is important. This also affects the VM resources used and price to be paid to run the application on the DSPS. We track the CPU and memory utilization for the dataflow as the average of the CPU and memory utilization across all the VMs that are being used by the dataflow's tasks. The per-VM information can also help identify which VMs hosting which tasks are the potential bottlenecks, and can benefit from data-parallel scale-out.

5   Proposed Benchmarks and Workload

We propose IoT benchmark workloads to help evaluate the metrics discussed before for various DSPS. The benchmarks have two parts: the dataflow logic that is executed on the DSPS and the input data streams that they consume.

5.1   IoT Input Stream Workloads

Sense your City (CITY) [1]. This is an *urban environmental monitoring* project[3] that crowd-sourced deployment of sensors at 7 cities across 3 continents in 2015, with about 12 sensors per city. Five timestamped observations: temperature, humidity, ambient light, dust and air quality, are reported every minute by a sensor along with the sensor ID and location. Besides urban sensing, this real-world data also captures the vagaries crowd-sourcing for IoT (Table 1).

Table 1. Smart Cities data stream features and rates at 1000× scaling

Dataset	Attributes	Format	Size (bytes)	Peak rate (msg/sec)	Distribution
CITY [1]	9	CSV	100	7,000	Normal
TAXI [11]	10	CSV	191	4,000	Bimodal

We use a single logical stream that combines the data from all 90 sensors. Since practical deployments of environmental sensing can easily extend to thousands of sensors per city, we use a temporal scaling of 1000× the native input rate to simulate a deployment of 90,000 sensors. Figure 2a shows a narrow normal distribution of the event rate centered at 6,400 msg/sec with a peak of 7,000 msg/sec. We use 7 days of data from 27 Jan to 2 Feb, 2015 for our benchmark.

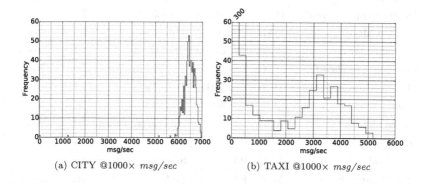

(a) CITY @1000× *msg/sec*   (b) TAXI @1000× *msg/sec*

Fig. 2. Frequency distribution of input throughputs for CITY and TAXI streams at 1000× temporal scaling used for the benchmark runs.

NYC Taxi cab (TAXI) [11]. This offers a stream of *smart transportation* messages that arrive from 2*M* trips taken in 2013 on 20,355 New York city taxis equipped with GPS[4]. A message is generated when a taxi completes a

---

[3] http://map.datacanvas.org.
[4] http://www.debs2015.org/call-grand-challenge.html/.

trip, and provides the taxi and license details, the start and end coordinates and timestamp, the distance traveled, and the cost, including the taxes and tolls.

Considering that events may be generated from the GPS sensors periodically rather than only at the end of the trip, we use a temporal scaling factor of $1000\times$ for our workload. This data has a bi-modal event rate distribution that reflects the morning and evening commutes, with peaks at 300 and $3,200$ events/sec. We use 7 days of data from 14-Jan-2013 to 20-Jan-2013 for our benchmark runs.

5.2   IoT Micro-benchmarks

We propose a suite of common micro-benchmark tasks that span various IoT categories and types of streaming task patterns as well. Their goal is to evaluate the performance of the DSPS for individual IoT tasks, using the *peak input throughput* that they can sustain on a unit computing resource as the performance measure. This offers a baseline for comparison with other DSPS, as well as when these tasks are used in application benchmarks with variable input rates (Table 2).

Table 2. IoT micro-benchmark tasks with different IoT categories and DSPS patterns

Task name	Code	Category	Pattern	σ ratio	State
XML parsing	XML	Parse	Transform	1:1	No
Bloom filter	BLF	Filter	Filter	1:0/1	No
Average	AVG	Statistical	Aggregate	N:1	Yes
Distinct appox. count	DAC	Statistical	Transform	1:1	Yes
Kalman filter	KAL	Statistical	Transform	1:1	Yes
Second order moment	SOM	Statistical	Transform	1:1	Yes
Decision tree classify	DTC	Predictive	Transform	1:1	No
Multi-variate linear reg.	MLR	Predictive	Transform	1:1	No
Sliding linear regression	SLR	Predictive	Flat map	N:M	Yes
Azure blob D/L	ABD	IO	Source/transform	1:1	No
Azure blob U/L	ABU	IO	Sink	1:1	No
Azure table query	ATQ	IO	Source/transform	1:1	No
MQTT publish	MQP	IO	Sink	1:1	No

We include a single XML parser as a representative parsing operation within our suite. The Bloom filter is a more practical filter operation for large discrete datasets, and we prefer that to a simple value range filter. We have several statistical analytics and aggregation tasks. These span simple averaging over a single attribute value to and second order moments over time-series values, to Kalman filter for denoising of sensor data and approximate count of distinct values for large discrete attribute values.

Predictive analytics using a multi-variate linear regression model that is trained offline and a sliding window univariate model that is trained online are included. A decision tree machine learning for discrete attribute values is also used for classification, based on offline training. Lastly, we have several IO tasks for reading and writing to Cloud file and NoSQL storage, and to publish to an MQTT publish-subscribe broker for notifications. We see that these tasks capture different dataflow patterns like transform, filter, aggregate and flat map.

5.3   IoT Application Benchmarks

Application benchmarks are valuable in understanding how non-trivial and meaningful IoT applications behave on DSPS. Application dataflows for a domain are most representative when they are constructed based on real or realistic application logic, rather than synthetic tasks. In case applications use highly-custom logic or proprietary libraries, this may not be feasible or reusable as a community benchmark. However, many of the common IoT tasks we have proposed earlier are naturally composable into application benchmarks that satisfy the requirements of a OODA decision making loop.

We propose application benchmarks that capture two common IoT scenarios: a *Data pre-processing and Statistical summarization (STATS)* application and a *Predictive Analytics (PRED)* application. STATS (Fig. 3a) ingests incoming data streams, performs data filtering of outliers on individual observation types using a Bloom filter, and then does three concurrent types of statistical analytics on observations from individual sensor/taxi IDs: sliding Average over a 90/10 event window for CITY/TAXI (~15 min native time window), Kalman filter for smoothing followed by a sliding window linear regression, and an approximate count of distinct readings. The outcomes from these statistics are published by an MQTT task, which can separately be subscribed to and visualized on a client.

The PRED dataflow captures the lifecycle of online prediction and classification to drive visualization and decision making for IoT applications. It parses incoming messages and forks it to a decision tree classifier and a multi-variate regression task. The decision tree uses a trained model to classify messages into classes, such as good, average or poor air quality, based on one or more of their attribute values. The linear regression uses a trained model to predict an attribute value in the message using several others. It then estimates the error $\frac{|p - o|}{o}$ between the predicted and observed value, normalized by the sliding average of the observations. These outputs are then grouped and plotted, and the file written to Cloud storage for hosting on a portal. One realistic addition is the use of a separate stream to periodically download newly trained classification and regression models from Cloud storage, and push them to the prediction tasks.

As such, these applications leverage many of the compositional capabilities of DSPS. The dataflows include *single and dual sources,* tasks that are *composed sequentially and in parallel, stateful and stateless* tasks, and *data parallel tasks* allowing for concurrent instances. The initial parse task for STATS uses a *flat map* pattern to create observation-specific streams. These are further grouped by their observation type using a *hash pattern* and passed to task instances.

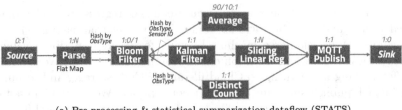

(a) Pre-processing & statistical summarization dataflow (STATS)

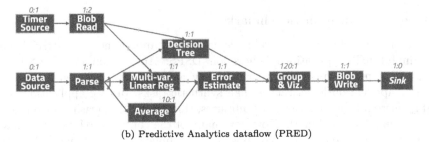

(b) Predictive Analytics dataflow (PRED)

Fig. 3. Application benchmarks composed using the micro-benchmark tasks.

6   Evaluation of Proposed Benchmarks

We implement the 13 micro-benchmarks as generic Java tasks that can consume and produce objects[5]. We validate our proposed benchmark by composing and running these dataflows on the popular Apache Storm open source DSPS.

In Storm, each task logic is wrapped by a *bolt* that invokes the task for each incoming tuple and emits response tuples. The dataflow is composed as a *topology* that defines the edges between the bolts, and the *groupings* which determine duplicate or hash semantics. We have implemented a scalable source task (*spout*) that replays events from a CSV file with a scaling factor. We generate random integers as tuples at a constant peak rate for the micro-benchmarks, and replay the original CITY and TAXI datasets at 1000× scaling for the applications.

We use Apache Storm 1.0.0 running on OpenJDK 1.7 and CentOS, and hosted on Microsoft Azure Cloud Virtual Machines (VMs). For the micro-benchmarks, Storm runs the benchmark task on one exclusive D1 VM (1-core Intel Xeon E5@2.2 GHz, 3.5 GiB RAM, 50 GiB SSD), while the source and sink tasks and the master service run on a D8 VM (8-core Intel Xeon E5@2.2 GHz, 28 GiB RAM, 400 GiB SSD). The large VM for the supporting services ensures that they are not the bottleneck when benchmarking the peak task rate on 1 VM. For the STATS and PRED application benchmark, we use D8 VMs for all the tasks of the dataflow, while reserving additional D8 VMs to exclusively run the supporting service. Each experiment runs for ∼10 min, which translates to about 7 days of event data for the CITY and TAXI datasets at 1000× scaling[6].

---

[5] https://github.com/dream-lab/bm-iot.

[6] Application runtime = $\frac{7\,\text{days} \times 24\,\text{h} \times 60\,\text{min} \times 60\,\text{s}}{1000 \times scaling}\,secs = 10.08\,\text{min}$.

6.1   Micro-benchmark Results

Figure 4 shows plots of the different metrics evaluated for the micro-benchmark tasks on Storm when running at their peak input rate supported on a single D1 VM with one thread. The *peak sustained throughput* per task is shown in Fig. 4a in *log-scale*. We see that most tasks can support $3,000$ msg/sec or higher rate, going up to $68,000$ msg/sec for BLF, DAC, KAL, DTC and MLR. XML parsing is highly CPU bound and has a peak throughput of only 310 msg/sec, and the Azure operations are I/O bound on the Cloud service and even slower.

The inverse of the peak sustained throughput gives the *mean latency*. However, it is interesting to examine the *end-to-end latency*, calculated as the time taken between emitting a message from the source, having it pass through the benchmarked task, and arrive at the sink task. This is the effective time contributed to the total tuple latency by this task running within Storm, including framework overheads. We see that while the mean latencies should be in sub-milliseconds for the observed throughputs, the box plot for end-to-end latency (Fig. 4b) varies widely up to $2,600$ ms for Q3. This wide variability could be because of non-uniform task execution times due to which slow executions queue up incoming tuples that suffer higher queuing time, such as for DTC and MLR that both use the WEKA library. Or tasks supporting a high input rate in the order of $10,000$ msg/sec, such as DAC and KAL, may be more sensitive to even small per-tuple overhead of the framework, say, caused by thread contention between the Storm system and worker threads, or queue synchronization. The Azure tasks that have a lower throughput also have a higher end-to-end latency, but much of which is attributable directly to the task latency.

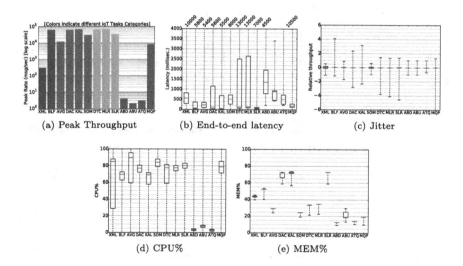

Fig. 4. Performance of micro-benchmark tasks for integer input stream at peak rate.

The box-plot for *jitter* (Fig. 4c) shows values close to zero in all cases. This indicates the long-term stability of Storm in processing the tasks at peak rate, without unsustainable queuing of input messages. The wider whiskers indicate the occasional mismatch between the expected and observed output rates.

The box plots for CPU utilization (Fig. 4d) shows the single-core VM effectively used at 70% or above in all cases except for the Azure tasks that are I/O bound. The memory utilization (Fig. 4e) appears to be higher for tasks that support a high throughput, potentially indicating the memory consumed by messages waiting in queue rather than consumed by the task logic itself.

6.2   Application Results

The STATS and PRED application benchmarks are run for the CITY and TAXI workloads at 1000× their native rates, and the performance plots shown in Fig. 5. The end-to-end latencies of the applications depend on the sum of the end-to-end latencies of each task in the critical path of the dataflow. The peak rates supported by the tasks in STATS is much higher than the input rates of CITY and TAXI. So the latency box plot for STATS is tightly bound (Fig. 5a) and its median much lower at 20 ms compared to the end-to-end latency of the tasks at their peak rates. The jitter is also close to zero in all cases. So Storm can comfortably support STATS for CITY and TAXI on 7 and 5 VMs, respectively. The distribution of VM CPU utilization is also modest for STATS. CITY has a 35% median with a narrow box (Fig. 5d), while TAXI has a low 5% median

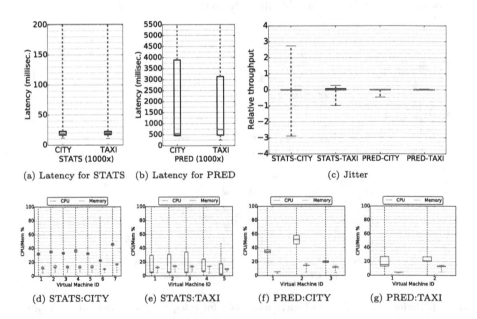

(a) Latency for STATS   (b) Latency for PRED   (c) Jitter

(d) STATS:CITY   (e) STATS:TAXI   (f) PRED:CITY   (g) PRED:TAXI

Fig. 5. End-to-end latency and Jitter (top), and CPU and Memory utilization (bottom) plots for STATS and PRED application benchmarks on CITY and TAXI workloads.

with a wide box (Fig. 5e) – this is due to its bi-modal distribution with low input rates, hence utilization, at nights, and high rates and utilization in the day.

For PRED, we see that the latency box plot is much wider, and the median end-to-end latency is between 500–700 ms for CITY and TAXI (Fig. 5b). This reflects the variability in task execution times for the WEKA tasks, DTC and MLR, which was observed in the micro-benchmarks too. The Azure blob upload also adds to the absolute increase in the end-to-end time. The jitter however remains close to zero, indicating sustainable performance. The CPU utilization is also higher, reflecting its more complex task logic relative to STATS.

7 Conclusion

In this paper, we have proposed a novel application benchmark for evaluating DSPS for IoT domains. These help evaluate common IoT tasks, as well as fully-functional applications for summarization and predictive analytics using with two real-world workloads from smart cities. The benchmark has been validated for the popular Apache Storm DSPS, and the performance metrics presented.

Acknowledgement. This work was supported by grants from the Robert Bosch Center for Cyber Physical Systems (RBCCPS) at IISc, DeitY and Microsoft Azure.

References

1. Data Canvas Dataset. http://datacanvas.org/sense-your-city/
2. Apache Flink. https://flink.apache.org/features.html/, April 2015
3. Agrawal, D., et al.: SparkBench – a spark performance testing suite. In: Nambiar, R., Poess, M. (eds.) TPCTC 2015. LNCS, vol. 9508, pp. 26–44. Springer, Cham (2016). doi:10.1007/978-3-319-31409-9_3
4. Ali, M.I., Gao, F., Mileo, A.: CityBench: a configurable benchmark to evaluate RSP engines using smart city datasets. In: Arenas, M., et al. (eds.) ISWC 2015. LNCS, vol. 9367, pp. 374–389. Springer, Heidelberg (2015). doi:10.1007/978-3-319-25010-6_25
5. Aman, S., Simmhan, Y., Prasanna, V.K.: Holistic measures for evaluating prediction models in smart grids. IEEE TKDE **27**(2), 475–488 (2015)
6. Arasu, A., Cherniack, M., Galvez, E., Maier, D., Maskey, A.S., Ryvkina, E., Stonebraker, M., Tibbetts, R.: Linear road: a stream data management benchmark. In: VLDB (2004)
7. Arlitt, M., Marwah, M., Bellala, G., Shah, A., Healey, J., Vandiver, B.: IoTAbench: an internet of things analytics benchmark. In: ICPE (2015)
8. Balazinska, M., Balakrishnan, H., Madden, S.R., Stonebraker, M.: Fault-tolerance in the borealis distributed stream processing system. ACM TODS (2008)
9. Chen, J., DeWitt, D.J., Tian, F., Wang, Y.: Niagaracq: a scalable continuous query system for internet databases. ACM SIGMOD Rec. **29**(2), 379–390 (2000)
10. Cooper, B.F., Silberstein, A., Tam, E., Ramakrishnan, R., Sears, R.: Benchmarking cloud serving systems with YCSB. In: ACM SoCC, pp. 143–154. ACM (2010)
11. Donovan, B., Work, D.B.: Using coarse GPS data to quantify city-scale transportation system resilience to extreme events. In: Transportation Research Board 94th Annual Meeting (2014)

12. Ghazal, A., Rabl, T., Hu, M., Raab, F., Poess, M., Crolotte, A., Jacobsen, H.A.: Bigbench: towards an industry standard benchmark for big data analytics. In: ACM SIGMOD (2013)

13. Huang, S., Huang, J., Dai, J., Xie, T., Huang, B.: The Hibench benchmark suite: characterization of the MapReduce-based data analysis. In: IEEE ICDEW (2010)

14. Lu, R., Wu, G., Xie, B., Hu, J.: Stream bench: towards benchmarking modern distributed stream computing frameworks. In: IEEE/ACM UCC (2014)

15. Nabi, Z., Bouillet, E., Bainbridge, A., Thomas, C.: Of streams and storms. Technical report, IBM (2014)

16. Nambiar, R.O., Poess, M.: The making of TPC-DS. In: VLDB (2006)

17. Suhothayan, S., Gajasinghe, K., Loku Narangoda, I., Chaturanga, S., Perera, S., Nanayakkara, V.: Siddhi: a second look at complex event processing architectures. In: ACM Workshop on Gateway Computing Environments (2011)

18. Toshniwal, A., Taneja, S., Shukla, A., Ramasamy, K., Patel, J.M., Kulkarni, S., Jackson, J., Gade, K., Fu, M., Donham, J., et al.: Storm@ twitter. In: ACM SIGMOD, pp. 147–156 (2014)

19. Wolf, G.: The data-driven life. The New York Times Magazine (2010)

20. Zaharia, M., Das, T., Li, H., Shenker, S., Stoica, I.: Discretized streams: an efficient and fault-tolerant model for stream processing on large clusters. In: USENIX Hot Cloud (2012)

AdBench: A Complete Benchmark for Modern Data Pipelines

Milind Bhandarkar$^{(\boxtimes)}$

Ampool Inc., Santa Clara, USA
milind@ampool.io

Abstract. Since the introduction of Apache YARN, which modularly separated resource management and scheduling from the distributed programming frameworks, a multitude of YARN-native computation frameworks have been developed. These frameworks specialize in specific analytics variants. In addition to traditional batch-oriented computations (e.g. MapReduce, Apache Hive [14] and Apache Pig [18]), the Apache Hadoop ecosystem now contains streaming analytics frameworks (e.g. Apache Apex [8]), MPP SQL engines (e.g. Apache Trafodion [20], Apache Impala [15], and Apache HAWQ [12]), OLAP cubing frameworks (e.g. Apache Kylin [17]), frameworks suitable for iterative machine learning (e.g. Apache Spark [19] and Apache Flink [10]), and graph processing (e.g. GraphX). With emergence of Hadoop Distributed File System and its various implementations as preferred method of constructing a *data lake*, end-to-end data pipelines are increasingly being built on the Hadoop-based data lake platform.

While benchmarks have been developed for individual tasks, such as Sort (TPCx-HS [5]), and Analytical SQL queries (TPC-xBB [6]), there is a need for a standard benchmark that exercises various phases of an end-to-end data pipeline in a data lake. In this paper, we propose a benchmark called *AdBench*, which combines Ad-Serving, Streaming Analytics on Ad-serving logs, streaming ingestion and updates of various data entities, batch-oriented analytics (e.g. for Billing), Ad-Hoc analytical queries, and Machine learning for Ad targeting. While this benchmark is specific to modern Web or Mobile advertising companies and exchanges, the workload characteristics are found in many verticals, such as Internet of Things (IoT), financial services, retail, and healthcare. We also propose a set of metrics to be measured for each phase of the pipeline, and various scale factors of the benchmark.

1 Introduction

As we witness the rapid transformation in data architecture, where RDBMS is being supplemented by large scale non-Relational stores, such as HDFS [11], MongoDB [23], Apache Cassandra [9], and Apache HBase [13], a more fundamental shift is on its way, which would require larger changes to modern data architectures. While the current shift was mandated by business requirements for the connected world, the next wave will be dictated by operational cost

© Springer International Publishing AG 2017
R. Nambiar and M. Poess (Eds.): TPCTC 2016, LNCS 10080, pp. 107–120, 2017.
DOI: 10.1007/978-3-319-54334-5_8

optimization, transformative changes in the underlying infrastructure technology, and newer use-cases, such as Internet of Things (IoT), deep learning, and conversational user interfaces (CUI).

These new use-cases combined with ubiquitous data characterized by its large volume, variety, and velocity has resulted in an emergence of big data phenomenon. Most applications being built in this connected world are based on analyzing historical data of various genres to create a context within which the user interaction with applications is interpreted. Such applications are popularly referred to as data-driven applications. While such data-driven applications are coming into existence across almost all the verticals, such as financial services, healthcare, manufacturing, and transportation, Web-based & mobile advertising companies recognized the potential of big data technologies for building data-driven applications, and were arguably the earliest adopters of big-data technologies. Indeed, popular big-data platforms such as Apache Hadoop [11] were created and adopted at web-technology companies, such as Yahoo, Facebook, Twitter, and Linkedin.

Most of the initial uses of these big data technologies were in workloads analyzing large unstructured and semistructured datasets in batch-oriented Java MapReduce programs, or using queries in SQL-like declarative languages. However, availability of majority of enterprise data on big data distributed storage infrastructure soon attracted other non-batch workloads to these platforms. Massively parallel SQL engines such as Apache Impala [15], and Apache HAWQ [12] that provide interactive analytical query capabilities were developed to natively run on top of Hadoop analyzing data stored in HDFS. Since 2012, cluster resource management, scheduling, and monitoring functionality previously embedded in MapReduce framework has been modularly separated out as a separate component, Apache YARN [11] in the Hadoop ecosystem. This allowed data processing frameworks other than MapReduce to natively run on Hadoop clusters, sharing computational resources among various frameworks. The last four years has seen an explosion of Hadoop-native data processing frameworks tackling various processing paradigms, such as streaming analytics, OLAP cubing, iterative machine learning, and graph processing, among others.

Flexibility of HDFS (and other Hadoop-compatible storage systems,) sometimes federated[1] across a single computational cluster, for storing large amounts of structured, semistructured, and unstructured data, and the added flexibility and choice of data processing frameworks, has resulted in Hadoop-based big data platforms that are being increasingly utilized for end-to-end data pipelines.

Emergence of this ubiquitous new platform architecture has created the imperative for an industry standard benchmark suite that evaluates efficacy of such platforms for end-to-end data processing workloads for consumers to make informed choices about various storage infrastructure components, and computational engines.

In this paper, we propose an end-to-end data pipeline benchmark, based on real-life workloads. While the benchmark itself is based on a typical pipeline in

---

[1] Referred to in the industry as a "Data Lake".

online advertising, it shares many common patterns across a variety of industries that make it relevant to a majority of big data deployments. In the next section, we discuss various previous benchmarking efforts, that have contributed to our current understanding of industry-standard big data benchmarks. In Sect. 3, we describe the benchmark scenario, at a fictitious Ad-Tech company. The datasets stored and manipulated in the end-to-end data pipeline of this benchmark are discussed in Sect. 4. In Sect. 5, we outline various computations that are performed on the data in data pipeline. We have developed a prototype implementation of the AdBench data pipeline that we describe in Sect. 6. In Sect. 7, we propose metrics to measure the effectiveness of systems under test (SUT), and report based on various classes of benchmarks that correspond to data sizes. We conclude in Sect. 8 after summarizing future work.

2   Related Work

Performance benchmarking of data processing systems is a rich field of study and has seen tremendous activity since commercial data platforms were first developed several decades ago. Several industry consortia have been established to standardize benchmarks for data platforms over the years. Of these, Transaction Processing Performance Council (TPC [7]) and Standard Performance Evaluation Corporation (SPEC [4]), both established in 1988, have gained prominence, and have created several standard benchmark suites. While TPC has traditionally been focused on data platforms, SPEC has produced several microbenchmarks aimed at evaluating low-level computer systems performance. Among the various TPC standard benchmarks, TPC-C (for OLTP workloads), TPC-H (for data warehousing workloads), and TPC-DS (for analytical decision support workloads) have become extremely popular, and several results have been published using these benchmark suites. The benchmark specifications consist of data models, tens of individual queries on these data, synthetic data generation utilities, and a test harness to run these benchmarks, and to measure performance.

Starting late 2011, the Center for Large-scale Data Systems Research (CLDS) at the San Diego Supercomputing Center, University of California at San Diego, along with several industry and academic partners, established a *Big Data Benchmarking Community (BDBC)* to establish standard benchmarks in evaluating hardware and software systems that form the modern big data platforms. A series of well-attended workshops called *Workshop on Big Data Benchmarking* have been conducted across three continents as part of this effort, and proceedings of those workshops have been published [24–26]. This author has been an active participant in formation of BDBC, and organizing the series of workshops. The AdBench proposal in this paper is a result of discussions and interactions within the BDBC and WBDB.

Recognizing the need to speed up development of industry standard benchmarks, especially for rapidly evolving big data platforms and workloads, a new approach to developing and adopting benchmarks within the TPC, called *TPC-express* [22] was adopted in 2014. Since then, two standard benchmarks, TPC-xHS [5], and TPC-xBB [6] have been created under TPC-express umbrella. TPC-xHS provides standard benchmarking rigor to the existing Hadoop Sort (based on Terasort) benchmark, whereas TPC-xBB provides big data extensions to the data models and workloads from the earlier TPC-DS benchmarks. We consider TPC-xHS to be a microbenchmark, which evaluates the shuffle performance in MapReduce and other large data processing frameworks. TPCx-BB is based on a benchmark proposal [1] called *BigBench* [21], that consists of a set of thirty modern data warehousing queries that combine both structured and unstructured data. None of the existing benchmarks provide a way to evaluate combined performance of an end-to-end data processing pipelines.

The need for having an industry-standard big data benchmark for evaluating performance of end-to-end data pipelines and conducting an annual competition to rank the top 100 systems based on this benchmark has been described in [2]. AdBench, the benchmark proposal in this paper builds upon this deep analytics pipeline benchmark proposal, by extending the proposed user-behavior deep-analytics pipeline, to include streaming analytics, and ad-hoc analytical queries, in the presence of transactions on the underlying datasets.

AdBench represents a generational advance over the previous standardized benchmarks, because it combines various common computational needs into a single end-to-end data pipeline. We note that while the performance of individual tasks in data pipelines remains important, operational complexity of assembling an entire pipeline composed of many different systems adds overheads of data exchange between various computational frameworks. Data layouts optimized for different tasks requires transformations. Scheduling different tasks into a single data pipeline requires a workflow engine with its own overheads. These additional data pipeline composition and orchestration is not taken into account by the previous benchmarks that measure effectiveness of systems on individual tasks. We believe that systems that exhibit good-enough performance to implement entire data pipelines are preferred for their operational simplicity, rather than best-of-breed disparate but incompatible systems that may exhibit the best performance for individual tasks, but have to be integrated with external glue code. Thus, our data pipeline benchmarks measures performance of an end-to-end data pipeline, rather than focus on *hero numbers* for individual tasks.

3   Benchmark Scenario

Acme is a very popular content aggregation company that has a web-based portal and also a mobile app, with tens of millions of users, who frequently visit them from multiple devices several times a day to get hyper-personalized content, such as news, photos, audio and video. Acme has several advertising customers[2] who

---

[2] We distinguish between *users*, who browse through Acme's website, from *customers*, who publish advertisements on that website.

pay them to display their advertisements on all devices. Acme is one of the many Web 3.0 companies because they have a deep understanding of their users' precise interests as well as exact demographic data, history of consumption of content, and interactions with advertisements displayed on Acme website and mobile application. Acme personalizes their users' experience based on this data. They have an ever growing taxonomy of their users' interests, and their advertisers can target users not only by the demographics, but also by the precise interests of those users.

Here is Acme's business in numbers:

- 100 Million registered users, with 50 Million daily unique users
- 100,000 advertisements across 10,000 advertisements campaigns
- 10 Million pieces of content (News, Photos, Audio, Video)
- 50,000 keywords in 50 topics & 500 subtopics as user interests, content topics, and for ad targeting

Acme has several hundred machines serving advertisements, using a unique matching algorithm, that fetches a user's interests, finds the best match within a few milliseconds, and serves the ad within appropriate content.

Acme has many data scientists, and Hadoop expertise that operates a large Hadoop cluster for providing personalized recommendations of content to their users. However, the batch-oriented nature of Hadoop has so far prevented them from using that Hadoop infrastructure for real-time ad serving, streaming analytics on advertisement logs, and providing real-time feedback on advertisement campaign performance to their customers. Also, for their business & marketing analysts, who want to perform ad-hoc queries on the advertising data to target a larger pool of users, they have set up a separate data repository away from both the real time analytics & batch analytics systems. As a result, the operating expenses of their data infrastructure have more than tripled. Worse, they incur a huge overhead, just trying to keep the data synchronized across these three platforms. Since essentially the same data is kept in multiple places, there is a lag & discrepancies in the data, and repetitive tasks for data cleansing, ensuring data quality, and maintaining data governance waste more than 80% of precious and valuable time of their data scientists, and big data infrastructure specialists.

Acme architects and engineers decided to replace the entire data infrastructure with modern flash & memory based architecture. However, because of the high costs of these modern hardware systems, they are unsure whether the return-on-investment will justify these increased capital expenditures. In addition, they need to consider the engineering costs of rewriting their entire existing data pipelines, written over more than five years, for these modern architectures. There are also considerations of having to retrain the data practitioners for utilizing new technologies. Based on process considerations, they decided to do incremental, piecemeal upgrades to their data infrastructure, moving towards a more unified data processing platform from their current disparate systems.

However, evaluating modern infrastructures, without having to rewrite the entire data pipeline, would require a standard benchmark, that is necessary for Acme to evaluate new systems.

4   The Data

There are four important large datasets used in Acme's data analysis pipeline.

4.1   User Profiles

User profiles contain details about every registered user. The schema for user profile is as follows:

- UserID: UUID (64–128 bit unique identifier.)
- Age: 0..255
- Sex: M/F/Unknown
- Location: Lat-Long
- Registration timestamp: TS
- Interests: Comma separated list of (topic:subtopic:keyword)

4.2   Advertisements

This dataset contains all the details about all the advertisements available for displaying within content. The schema for this dataset is as follows:

- AdID: UUID
- CampaignID: UUID
- CustomerID: UUID
- AdType: {"banner", "modal", "search", "video", "none"}
- AdPlatform: {"web", "mobile", "both"}
- Keywords: comma separated list of (topic:subtopic:keyword)
- PPC: $ per click
- PPM: $ per 1000 ads displayed
- PPB: $ per conversion

4.3   Content Metadata

Content dataset contains all the metadata about the content. The schema for the content dataset is as follows:

- ContentID: UUID
- ContentType: {"news", "video", "audio", "photo"}
- Keywords: Comma separated list of (topic:subtopic:keyword)

4.4   Ad Serving Logs

This dataset is streamed continuously from the ad servers. Each entry in this log has the following fields, of which some may be null:

– TimeStamp: TS
– IPAddress: IPv4/IPv6
– UserID: UUID
– AdID: UUID
– ContentID: UUID
– AdType: {"banner", "modal", "search", "video"}
– AdPlatform: {"web", "mobile"}
– EventType: {"View", "Click", "Conversion"}

5   Computations

Following computational steps are performed on the data in Acme's advertisement analytics data pipelines.

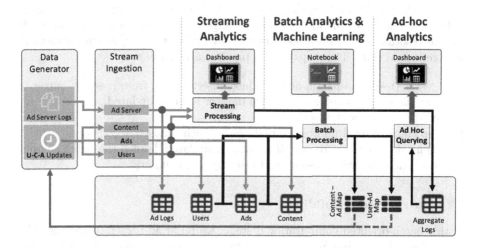

Fig. 1. Dataflow across AdBench data pipeline

5.1   Ingestion and Streaming Analytics

This phase of the data pipeline is based on a streaming analytics benchmark proposed by the Yahoo! Storm Engineering Team [27]. Ad servers produce ad click, view, & conversion events to a message queue, or a staging storage system. Queue consumers consume events continuously from a message queue, and process them in a streaming manner using a streaming analytics platform. For every event consumed the following computations are performed:

1. Parse the event record
2. Extract Timestamp, AdID, EventType and AdType
3. Look up CampaignID from AdID
4. Windowed aggregation of event types for each AdID, and CampaignID
5. Store these aggregates in an aggregate dataset
6. Prepare these aggregations for a streaming visualization dashboard for a CampaignID, and all Ads in that Campaign

The two output datasets from the streaming analytics stage is:

1. $(AdID, Window, nViews, nClicks, nCon, \sum PPV, \sum PPC, \sum PPCon)$
2. $(CmpgnID, Window, nViews, nClicks, nCon, \sum PPV, \sum PPC, \sum PPCon)$

Second streaming ingestion pipeline keeps the User table, Ad Table, and Content Table updated. While the ads are being displayed, clicked, and converted, new users are being registered, and existing users' information is being updated. New campaigns are created, existing campaigns are modified, new ads are being created, and existing ads are updated. The correct implementation of this data pipeline will allow these inserts and updates taking place concurrently with other stages of the pipeline, rather than periodically, thus introducing transactionality. In this ingestion pipeline also, we use message queue consumers to get insert & update records, and apply these inserts and updates to respective datasets in storage in real time. The steps in this pipeline are as follows:

1. Ingest a {user, campaign, ad} {update, insert} event from message queue.
2. Parse the event to determine which dataset is to be updated.
3. Update respective dataset.
4. Keep track of total number of updates for each dataset.
5. When 1% of the records are either new or updated, launch the batch computation stage described below, and reset update counters.

5.2   Batch Model Building

In this batch-oriented computational stage, we build ad targeting models. The inputs for this pipeline are the user dataset, ad dataset, and content dataset. And output of this pipeline are two new datasets:

1. $(UserID, AdID_1, weight_1, AdID_2, weight_2, AdID_3, weight_3)$
2. $(ContentID, AdID_1, weight_1, AdID_2, weight_2, AdID_3, weight_3)$

These datasets represent the top 3 most relevant ads for every user, and for every content. These datasets are then used for the Ad serving systems, such that when a user visits a particular content, the best match among these ads are chosen, based on one look up each in the user dataset, and content dataset.

Relevance of an Ad for a user or a content is determined by cosine similarity in the list of keywords, and topics and subtopics. This model building pipeline, built using batch-oriented computational frameworks, has the following steps:

1. Extract the relevant fields from user dataset & ad dataset, and join based on topics, subtopics and keywords.
2. Filter the top 3 matching keywords, and compute the weights of ads for those keywords using cosine similarity.
3. Repeat the steps above for the content dataset & ad dataset.

5.3   Interactive and Ad-Hoc SQL Queries

The interactive and ad-hoc queries are performed on varying windows of the aggregates for campaigns and ads using an interactive query engine (preferably SQL-based). Some examples of the queries are:

1. What was the per-minute, hourly, daily conversion rate for an Ad? For a campaign?
2. How many Ads were clicked on as a percentage of viewed, per hour for a campaign?
3. How much money does a campaign owe to Acme for the whole day?
4. What are the most clicked ads & campaigns per hour?
5. How many male users does Acme have aged 0–21, 21–40?

The results of these queries can be displayed in a terminal, and for the queries resulting in time-windowed data, visualized using BI tools.

Various stages of computations in the AdBench data pipeline, and dataflow among them is shown in Fig. 1.

6   Prototype Implementation

At Ampool, we are building a next-generation data platform that enables unified analytics on structured, semistructured, and unstructured data. At the core of our product is a distributed, memory-centric, highly available, elastic object store, that acts as a substrate for hybrid transactional and analytical processing (HTAP) workloads. In addition, we have connectors to common open source analytical computation frameworks.

We have built a prototype implementation of AdBench, where Ampool is used as a unified data store for the raw & derived datasets. For Streaming data ingestion, we have used Apache Kafka [16]. We simulate Ad servers with our synthetic data generator, which generates Ad Server Logs. Currently the synthetic data generator is only a single instance, simulating a single Ad server, but in future, we will extend it to simulate multiple Ad servers. For user, content, and advertising campaign updates, we have another synthetic data generator, which is based on real-world catalog of products to generate keywords, and other ad targetting features.

For streaming data analytics, we use Apache Apex [8]. Apache Apex is a Hadoop YARN native platform that unifies stream and batch processing. It processes big data in-motion in a way that is highly scalable, highly performant, fault tolerant, stateful, secure, distributed, and easily operable. All the operations, such as computing the windowed aggregations for advertisements served, clicked etc. are implemented as operators in the Apache Apex platform.

For batch data analytics, we use Cask data Application Platform (CDAP [3]). CDAP is an open source framework to build and deploy data applications on Apache Hadoop. CDAP is an abstraction layer on top of Hadoop and other open source infrastructure such as HBase, Hive, MapReduce & Spark that enables developers to rapidly build, and operations to easily manage, real-time and batch data applications. For our AdBench batch-oriented machine learning, we have used Apache Hive to do pre-aggregation of the raw advertisement logs, user preferences, and advertisement and content keywords. Based on these aggregation, we run recommendation algorithm written using Apache Spark, to generate weights for individual advertisements, and their relevance to the user preference. In the third stage of this pipeline, we take the top three best suited advertisements per user & piece of content, to create an ad serving dataset.

For interactive and ad-hoc queries, we use Apache Trafodion [20]. Apache Trafodion (incubating) is a webscale SQL-on-Hadoop solution enabling transactional or operational workloads on Apache Hadoop. Trafodion builds on the scalability, elasticity, and flexibility of Hadoop. Trafodion extends Hadoop to provide guaranteed transactional integrity, enabling new kinds of big data applications to run on Hadoop. Trafodion query-language is fully ANSI SQL compliant, with ODBC & JDBC support. We also use Trafodion's visual dashboard to generate visualizations from the above Ad-Hoc queries.

Figure 2 shows our prototype implementation of AdBench data pipeline using above mentioned technology components. We are planning to publish the source code, as well as our installation, deployment, and benchmark runner as open source project soon.

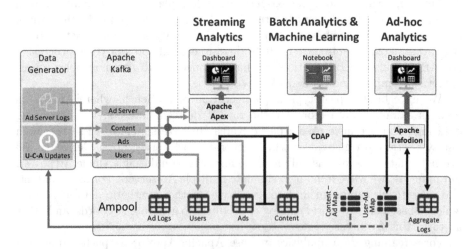

Fig. 2. Prototype implementation of AdBench data pipeline using ampool as data store

While the prototype implementation that uses several disparate computational frameworks on a single unified memory-based storage framework may seem complex, it is intended to emphasize the difficulties of integrating current tools to assemble these data processing pipelines. By highlighting this difficulty, we hope to guide development of unified computational frameworks, or additionally tools that simplify this integration. Indeed, we are witnessing limited unification of different computational engines on a single language runtime, which will simplify building the pipelines. For example, Apache Beam is a unified API with a modularly separated runtime, that allows convergence of streaming and batch-oriented analytics. Also, Apache Spark has support for SparkSQL for interactive and ad-hoc queries, Spark Streaming for streaming analytics, and Scala-based raw Spark interfaces to implement iterative machine learning computations. We expect AdBench and other end-to-end data pipeline benchmarks to be relevant to measure complexity of assembling such pipelines.

7   Scale Factors and Metrics

When compared to micro-benchmarks or benchmarks that consist of a fixed set of queries representing similar workloads, performed on a single system, an end-to-end data pipeline with mixed workloads poses significant challenges in defining scale factors, metrics, and reporting benchmark results. In this section, we propose a few alternatives, with reasoning behind our choice.

7.1   Scale Factors

Across different verticals, the number of entities and thus the amount of data, and data rates vary widely. For example, in financial institutions, such as Banks providing online banking, the number of customers (users) are between a few tens of thousands to millions, number of advertisements correspond to number of different services (such as loans, fixed deposits, various types of accounts) are tens to hundreds, number of content entities are equivalent to the number of dashboards (such as account transaction details, bill pay), and the data rates tend to be in low tens of thousands per minute even during peak times. These scales are very different than a content aggregation website or mobile application. To address these varying needs for various industries, we suggest the following scale factors:

Class	Users	Ads	Contents	Events/second
Tiny	100,000	10	10	1,000
Small	1,000,000	100	100	10,000
Medium	10,000,000	1,000	1,000	100,000
Large	100,000,000	10,000	10,000	1,000,000
Huge	1,000,000,000	100,000	100,000	10,000,000

7.2 Metrics

Since the benchmark covers an end-to-end data pipeline consisting of multiple stages, measures of performance in each of the stages vary widely. Streaming data ingestion is measured in terms of number of events consumed per second, but due to varying event sizes, amount of data ingested per second is also an important measure. Streaming data analytics operates on windows of event streams, and performance of streaming analytics should be measured in terms of window size (amount of time per window, which is a product of number of events per unit time, and data ingestion rate). Because of the latency involved in ingestion of data, and output of streaming analytics, the most relevant measure of performance of streaming analytics is the amount of time between event generation, and event participation in producing results.

Batch computations, such as machine-learned model training (of which recommendation engine is a representative example), amount of time needed from beginning of the model-training, including any preprocessing of raw data, to the upload of these models to the serving system needs to be considered as performance measure.

Similarly, for ad-hoc and interactive queries, time from query submission, to the production of results is the right performance measure.

While most of the benchmark suits attempt to combine the performance results into a single performance metric, we think that for an end-to-end data pipeline, one should report individual performance metrics of each stage, to enable choice of different components of the total technology stack. For a rapidly evolving big data technology landscape, where a complete data platform is constructed from a collection of interoperable components, we encourage experimentation of combining multiple technologies to implement end-to-end data pipeline, and stage-wise performance metrics will aid this decision.

In order to simplify comparisons of environments used to execute these end to end pipelines, initially, a simple weighted aggregation of individual stage's performance is proposed as a single performance metric for the entire system, with weights explicitly specified. There is anecdotal evidence to suggest that the business value of data decreases as time between production and consumption of data increases. Thus, one might be tempted to give higher weight to the streaming computation performance than to the batch computation performance. However, it also might be argued that approximate computations in real-time are acceptable, while batch computations need exact results, and therefore faster batch computations should be given higher weightage. We propose that determining appropriate weights for individual stages should be an active area of investigation, and should be incoporated in the metrics based on the outcome of this investigation.

In addition to performance metrics, the complexity of assembling data pipelines by combining multiple technologies must be taken into account by the architects of these pipelines. Skills needed to effectively use multiple technologies, and operational complexity of these data pipelines tend to be some of the important factors in choosing such systems. These are reflected in the total cost of ownership (TCO) of the system under test. We recommend that metrics such as time & and manpower required to develop the pipeline, time to deployment, management and monitoring tools for operationalizing data pipelines should be reported.

8 Conclusion

In this paper, we have proposed an end-to-end data pipeline to be used as a benchmark for evaluating various data and analytics platforms. While the particular usecase is specific to online advertisement technology industry, most industries have similar data pipeline workloads composed of multiple data processing stages. Therefore, we consider this as a representative benchmark. An end-to-end pipeline benchmark not only measures the performance of individual stages in the data pipeline, but also takes into account performance of data exchange and possible transformations between different stages. In future, we would like to extend this benchmark to include more data genres, such as graphical, and multi-media data.

References

1. Baru, C., et al.: Discussion of BigBench: a proposed industry standard performance benchmark for big data. In: Nambiar, R., Poess, M. (eds.) TPCTC 2014, vol. 8904, pp. 44–63. Springer, Heidelberg (2014)
2. Baru, C., Bhandarkar, M., Nambiar, R., Poess, M., Rabl, T.: Benchmarking big data systems and the bigdata top 100 list. Big Data 1(1), 60–64 (2013)
3. Cask Data, Inc., Cask Data Application Platform (CDAP), June 2016
4. Standard Performance Evaluation Corporation. SPEC Website, June 2016
5. Transaction Processing Performance Council. TPC Express Benchmark HS, Standard Specification, Version 1.4.0, April 2016
6. Transaction Processing Performance Council. TPC Express Big Bench, Standard Specification, Version 1.1.0, May 2016
7. Transaction Processing Performance Council. TPC Website, June 2016
8. Apache Software Foundation. Apache Apex, June 2016
9. Apache Software Foundation. Apache Cassandra, June 2016
10. Apache Software Foundation. Apache Flink, June 2016
11. Apache Software Foundation. Apache Hadoop, June 2016
12. Apache Software Foundation. Apache HAWQ (inbcubating), June 2016
13. Apache Software Foundation. Apache HBase, June 2016
14. Apache Software Foundation. Apache Hive, June 2016
15. Apache Software Foundation. Apache Impala, June 2016
16. Apache Software Foundation. Apache Kafka, June 2016
17. Apache Software Foundation. Apache Kylin, June 2016
18. Apache Software Foundation. Apache Pig, June 2016
19. Apache Software Foundation. Apache Spark, June 2016
20. Apache Software Foundation. Apache Trafodion (incubating), June 2016

21. Ghazal, A., Rabl, T., Hu, M., Raab, F., Poess, M., Crolotte, A., Jacobsen, H.-A.: Bigbench: towards an industry standard benchmark for big data analytics. In: Proceedings of the 2013 ACM SIGMOD International Conference on Management of Data, SIGMOD 2013, pp. 1197–1208. ACM, New York (2013)
22. Huppler, K., Johnson, D.: TPC express – a new path for TPC benchmarks. In: Nambiar, R., Poess, M. (eds.) TPCTC 2013. LNCS, vol. 8391, pp. 48–60. Springer, Heidelberg (2014). doi:10.1007/978-3-319-04936-6_4
23. MongoDB, Inc., MongoDB, June 2016
24. Rabl, T., Poess, M., Baru, C., Jacobsen, H.-A. (eds.): WBDB 2012. LNCS, vol. 8163. Springer, Heidelberg (2013)
25. Rabl, T., Jacobsen, H.-A., Raghunath, N., Poess, M., Bhandarkar, M., Baru, C. (eds.): WBDB 2013. LNCS, vol. 8585. Springer, Heidelberg (2014)
26. Rabl, T., Sachs, K., Poess, M., Baru, C., Jacobson, H.-A. (eds.): WBDB 2014. LNCS, vol. 8991. Springer, Heidelberg (2015)
27. Yahoo Storm Engineering Team. Benchmarking Streaming Computation Engines at Yahoo! December 2015

Lessons Learned: Performance Tuning
for Hadoop Systems

Manan Trivedi and Raghunath Nambiar[(✉)]

Cisco Systems, Inc., 275 East Tasman Drive, San Jose, CA 95134, USA
{matrived,rnambiar}@cisco.com

Abstract. Hadoop has become a strategic data platform for by mainstream enterprises, adopted because it offers one of the fastest paths for businesses take to unlock value from big data while building on existing investments. Hadoop is a distributed framework based on Java that is designed to work with applications implemented using MapReduce modeling. This distributed framework enables the platform to pass the load to thousands of nodes across the whole Hadoop cluster. The nature of distributed frameworks also allows node failure without cluster failure. The Hadoop market is predicted to grow at a compound annual growth rate (CAGR) over the next several years. Several tools and guides describe how to deploy Hadoop clusters, but very little documentation tells how to increase performance of Hadoop clusters after they are deployed. This document provides several BIOS, OS, Hadoop, and Java tunings that can increase the performance of Hadoop clusters. These tunings are based on lessons learned from Transaction Processing Performance Council Express (TPCx) Benchmark HS (TPCx-HS) testing on a Cisco UCS® Integrated Infrastructure for Big Data cluster. TPCx-HS is the industry's first standard for benchmarking big data systems. It was developed by TPC to provide verifiable performance, price-to-performance, and availability metrics for hardware and software systems that use big data.

Keywords: Hadoop · Tuning · Industry standard · TPCx-HS

1 Introduction

Big data is expected to fuel the next industrial revolution. An early sign is the wide adoption of big data technologies across major market sectors, including agriculture, education, entertainment, finance, healthcare, manufacturing, transportation, and government. According to IDC, the big data technology and services market experienced six times the growth rate of the overall information and communications technology market in 2015 [1]. This market is expected to be US$34 billion in 2017, and it is expected to be directly and indirectly responsible for US$300 billion in worldwide IT spending. This exponential growth in big data is fueled primarily by several open-source software initiatives and industry-standard infrastructure solutions.

The most prominent software platform by far is Hadoop. In fact, Hadoop and big data are often considered synonymous. Hadoop adaption is predicted to grow at a compound annual growth rate (CAGR) over the next several years across major industry

© Springer International Publishing AG 2017
R. Nambiar and M. Poess (Eds.): TPCTC 2016, LNCS 10080, pp. 121–141, 2017.
DOI: 10.1007/978-3-319-54334-5_9

vertical markets as a mainstream data management platform. Several tools and guides describe how to deploy Hadoop clusters, but very little documentation tells how to increase the performance of Hadoop clusters after they are deployed.

This document explains several BIOS, OS, Hadoop, and Java tunings that can increase the performance of Hadoop clusters. These tunings are based on lessons learned from Transaction Processing Performance Council Express (TPCx) Benchmark HS (TPCx-HS) testing. The tests were conducted on a Cisco UCS® Integrated Infrastructure for Big Data cluster, an industry-leading platform for enterprise Hadoop deployments. However, these tuning parameters are applicable across most Hadoop deployments.

This document also presents the results of tests addressing eight of the most frequently asked questions in tuning Hadoop systems. All test results reported are based on fully compliant TPCx-HS testing based on the specification, but they have not been audited or published.

2   TPC Express Benchmark HS

TPCx-HS is the industry's first standard for benchmarking big data systems. It is designed to provide verifiable performance, price-to-performance, and availability metrics for hardware and software systems that use big data [2, 3].

TPCx-HS can be used to assess a broad range of system topologies and implementation methodologies for Hadoop in a technically rigorous and directly comparable, vendor-neutral manner. And although modeling is based on a simple application, the results are highly relevant to big data hardware and software systems.

TPCx-HS benchmarking has three steps:

- HSGen: Generates data and retains it on a durable medium with three-way replication
- HSSort: Samples the input data, sorts the data, and retains the data on a durable medium with three-way replication
- HSValidate: Verifies the cardinality, size, and replication factor of the generated data

The TPCx-HS specification mandates two consecutive runs to demonstrate repeatability, as depicted in Fig. 1, and the lower value is used for reporting [4].

TPCx-HS uses three main metrics:

- HSph@SF: Composite performance metric, reflecting TPCx-HS throughput, where SF is the scale factor
- $/HSph@SF: Price-to-performance metric
- System availability date

TPCx-HS also reports the following numerical quantities:

- T_G: Data generation phase completion time, with HSGen reported in hh:mm:ss format
- T_S: Data sort phase completion time, with HSSort reported in hh:mm:ss format
- T_V: Data validation phase completion time, reported in hh:mm:ss format

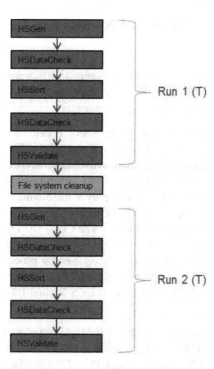

Fig. 1. TPCx-HS benchmark processing

The primary performance metric of the benchmark is HSph@SF, the effective sort throughput of the benchmarked configuration. Here is an example (using the summation method):

$$HSph@SF = \left\lfloor \frac{SF}{(T/3600)} \right\rfloor$$

Here, SF is the scale factor, and T is the total elapsed time for the run in seconds. The price-to-performance metric for the benchmark is defined as follows:

$$\$/HSph@SF = \frac{P}{HSph@SF}$$

Here, P is the total cost of ownership (TCO) of the system under test (SUT).

The system availability date indicates when the system under test is generally available as defined in the TPC-Pricing specification.

3   System Under Test: Cisco UCS Integrated Infrastructure for Big Data

The tests were conducted on a Cisco UCS Integrated Infrastructure for Big Data cluster with 16 Cisco UCS C240 M4 Rack Servers. The Cisco UCS Integrated Infrastructure for Big Data is built using the following components:

- Cisco UCS 6296UP 96-Port Fabric Interconnect: Fabric interconnects are central to the Cisco Unified Computing System™ (Cisco UCS). They provide low-latency, lossless 10 Gigabit Ethernet, Fibre Channel over Ethernet (FCoE), and Fibre Channel functions with management capabilities for the system. All servers attached to fabric interconnects become part of a single, highly available management domain.
- Cisco UCS C240 M4 Rack Server: Cisco UCS C-Series Rack Servers extend Cisco UCS in standard rack-mount form factors. The Cisco UCS C240 M4 Rack Server is designed to support a wide range of computing, I/O, and storage-capacity demands in a compact design. It supports two Intel® Xeon® processor E5-2600 v4 series CPUs, up to 768 GB of memory, and 24 small-form-factor (SFF) disk drives plus two internal SATA boot drives and Cisco UCS Virtual Interface Card (VIC) 1227 adapters.

The Cisco UCS Integrated Infrastructure for Big Data cluster configuration consists of two Cisco UCS 6296UP fabric interconnects, 16 Cisco UCS C240 M4 servers with two Intel Xeon processor E5-2600 v4 series CPUs, 256 GB of memory, and 24 SFF disk drives plus two internal SATA boot drives and Cisco UCS VIC 1227 adapters, as shown in Fig. 2. Table 1 lists the software versions used.

16 x Cisco UCS C240 M4 Servers (Data Nodes) with:
24 × 1.2-TB 6-Gbps SAS 10,000-rpm SFF HDD
2 × 120-GB 2.5-Inch Enterprise Value 6-Gbps SATA SSD (Boot)
10 Gigabit Ethernet
16 × 10 Gigabit Ethernet
2 x Cisco UCS 6296UP fabric interconnect
1 x Cisco Nexus® 9372PX Switch

Table 1. Software versions

Layer	Component	Version or Release
Computing	Cisco UCS C240 M4 server	Release C240M4.2.0.10c
Network	Cisco UCS 6296UP fabric interconnect	Release UCS 3.1(1 g)A
	Cisco UCS VIC 1227 firmware	Release 4.1(1d)
	Cisco UCS VIC 1227 driver	Release 2.3.0.18
Software	Red Hat Enterprise Linux (RHEL) server	Version 6.5 (x86_64)
	Cisco UCS Manager	Release 3.1(1 g)
Hadoop	Cloudera Enterprise	Version 5.3.2

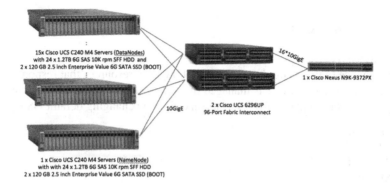

Fig. 2. Cisco UCS integrated infrastructure for big data cluster configuration

4   Performance Tuning

Many factors come into play when tuning a system as complex as big data systems. Performance tuning involves making modifications to hardware, software, and network parameters.

This section lists parameters that can be tuned at the infrastructure, operating system, and Hadoop levels.

Infrastructure

Infrastructure tuning helps achieve optimal utilization of resources. It also helps the application run faster and perform better.

- Server
  - BIOS
    CPU parameters
    Intel Turbo Boost Technology
    Intel Hyper-Threading Technology
    Prefetcher
    C-states
    Power control policy
    Memory tuning
- Network
  - Network tuning parameters
  - Network interface card (NIC) bonding
  - Jumbo frame (maximum transmission unit [MTU])
  - Quality-of-service (QoS) settings
- Storage
  - RAID 0
    Write back
    Read ahead
    Stripe size

- JBOD
- JBOD Versus RAID 0

Operating System

OS performance tuning is used to manage and improve resources that respond to individual requests. OS scalability is managed by monitoring the resource consumption of varying volumes of requests, from low to very high, by changing default OS settings.

- File system
  - XFS
  - Agcount
  - Mount
  - Fstab
- Post-OS tuning
  - sysctl.conf
  - limits.conf
  - CPU frequency and scaling governor
  - Transparent huge pages
  - Linux swappiness
  - I/O scheduler

Hadoop

In addition to tuning the infrastructure and OS, you need to tune Hadoop settings for best performance. Hadoop tuning can have a significant impact on the overall performance of your Hadoop cluster.

- Hadoop
  - Hadoop Distributed File System (HDFS)
       hdfs-site.xml
  - MapReduce
       Java Virtual Machine (JVM) reuse
       Compression
       mapred-site.xml
       core-site.xml

5   Performance Tuning in Detail

This section describes the infrastructure, OS, and Hadoop tuning parameters in detail.

Server Tuning

Hadoop is based on a new approach to storing and processing complex data, with data movement reduced. Hadoop distributes across the cluster the data that each machine in a Hadoop cluster stores, and it also processes the data. Therefore, it is important to tune the processing, or computing, aspect of the system to achieve optimal performance from the cluster.

BIToS settings can have a significant performance impact, depending on the workload and the applications. Table 2 lists the optimal CPU settings for Hadoop based on the tests reported in this document.

Table 2. Optimal CPU settings

Parameter	Setting
Intel Turbo Boost	Enabled
Enhanced Intel SpeedStep	Enabled
Intel Hyper-Threading	Enabled
Core Multiprocessing	All
Executive Disabled Bit	Platform default
Virtualization Technology	Disabled
Hardware Prefetcher	Enabled
Adjacent Cache Line Prefetcher	Enabled
Data Cache Unit (DCU) Streamer Prefetcher	Enabled
DCU IP Prefetcher	Enabled
Direct Cache Access	Enabled
Processor C-State	Disabled
Processor C1E	Disabled
Processor C3 Report	Disabled
Processor C6 Report	Disabled
Processor C7 Report	Disabled
CPU Performance	Enterprise
Maximum Variable Mean Time to Replace or Repair (MTRR) Setting	Platform default
Local x2APIC Advanced Programmable Interrupt Controller	Platform default
Power Technology	Performance
Energy Performance	Performance
Frequency Floor Override	Enabled
P-State Coordination	Hw-all
DRAM Clock Throttling	Performance
Channel Interleaving	Platform default
Rank Interleaving	Platform default
Demand Scrub	Disabled
Patrol Scrub	Disabled
Altitude	Platform default
Package C-State Limit	Platform default

Table 3 lists optimal memory settings for Hadoop based on the tests reported here.

Table 3. Optimal memory settings for Hadoop

Parameter	Setting
Memory RAS Configuration	Maximum performance
NUMA	Enabled
Low-Voltage Double Data Rate (LV DDR) Mode	Performance mode
DRAM Refresh Rate	1 time
DDR3 Voltage Selection	Platform default

Table 4. Optimal network tuning parameters for Hadoop

Parameter	Tuned value	Description
net.core.somaxconn	1024	Changing the net.core.somaxconn Linux kernel settings from the default of 128 to 1024 helps with burst requests from the name node and job tracker. This option sets the size of the listening queue, or the number of connections that the server can set up at one time.
net.ipv4.tcp_retries2	5	This variable helps forward the packets between interfaces. This variable is special; its change resets all configuration parameters to their default state.
net.ipv4.ip_forward	0	IP forwarding is disabled in most Linux distributions because most of them do not set up a Linux router, gateway, VPN server, or dial-in server.
net.ipv4.conf.default.rp_filter	1	This value influences the timeout behavior of a live TCP connection.
net.ipv4.conf.all.rp_filter	1	This value enables route verification on all interfaces.
net.ipv4.conf.default.accept_source_route	0	This setting does not accept source routing.
net.ipv4.tcp_syncookies	1	This setting enables the use of TCP SYN cookies.
net.ipv4.conf.all.arp_filter	1	
net.ipv4.tcp_mtu_probing	1	If there are multiple network interfaces on different IP addresses, this setting will help achieve the desired results.
net.ipv4.icmp_echo_ignore_broadcasts	1	This setting controls TCP packetization layer path MTU discovery. It is disabled by default, and it is enabled when an Internet Control Message Protocol (ICMP) black hole is detected.
net.ipv4.conf.default.promote_secondaries	1	These settings prevent deletion of secondary IP addresses when the primary IP address is deleted.
net.ipv4.conf.all.promote_secondaries	1	
net.core.rmem_max	16777216	These settings increase the TCP maximum buffer size. The four options shown here increase the TCP send and receive buffers, allowing an application to move its data out faster so it can serve other requests. This adjustment also improves the client's ability to send data to the server when it gets busy.
net.core.wmem_max	16777216	
net.ipv4.tcp_rmem	4096 87380 16777216	
net.ipv4.tcp_wmem	4096 65536 16777216	
net.core.netdev_max_backlog	10000	
net.core.netdev_max_backlog	10000	

Network Tuning

The impact of the network on big data is enormous. An efficient and resilient network is a crucial part of a good Hadoop cluster because the network is what connects all the nodes. The network is also crucial for writing data, reading data, and signaling and for HDFS operations and operations of the MapReduce infrastructure. Therefore, the failure of a networking device can have dire affects. A job may need to be restarted, or a workload may be pushed to the remaining nodes, resulting in delay. Therefore, networks must be designed to provide redundancy, with multiple paths between computing nodes, and they must be able to scale.

Table 4 lists some network performance settings that can increase Hadoop performance. These options increase the read and write cache sizes for the network stack. These parameters can be tested with the **systctl −w** command or made permanent by adding the variable to the /etc./sysctl.conf file.

You can tune NIC bonding. A NIC is a computer hardware component that connects a computer to a computer network. Network bonding is a method of combining (joining) two or more network interfaces together into a single interface. This combination increases network throughput and provides redundancy. If one interface is down or unplugged, the remaining interfaces will keep the network traffic up and alive. Network bonding can be used in situations in which you need redundancy, fault tolerance, or load balancing.

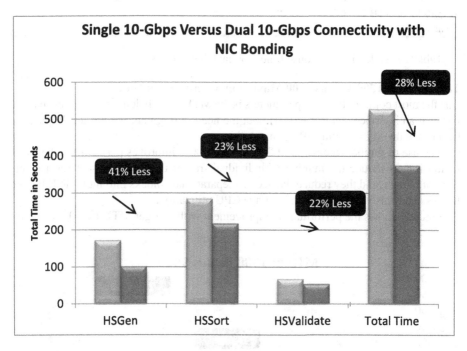

Fig. 3. **Single** 10-Gbps Versus **Dual 10**-Gbps Connectivity with NIC Bonding

Linux allows bonding of multiple network interfaces into a single channel using a special kernel module called a bonding module. The Linux bonding driver provides a method for aggregating multiple network interfaces into a single logical "bonded" interface. The behavior of the bonded interface depends on the mode. In general, the mode provides either hot-standby or load-balancing services. Additionally, link-integrity monitoring can be performed.

Test Result 1: 10-Gbps Versus Dual 10-Gbps Connectivity with NIC Bonding
One frequently asked question relates to the impact of NIC bonding for Hadoop. In older-generation servers, single 10-Gbps connectivity was sufficient. Since the introduction of Cisco UCS C240 M4 servers (based on Intel Xeon processor 2600 v3 CPUs) with 24 SFF disks drives, we have observed significant performance improvements with NIC bonding. In other words, Hadoop nodes can use more than 10-Gbps network bandwidth (Fig. 3).

Table 5. **Single** 10-Gbps versus **Dual 10**-Gbps **with NIC Bonding**

Phase	No Bonding (Time in Seconds)	2-NIC Bonding (Time in Seconds)	Percentage improvement
HSGen	173	102	41.0%
HSSort	286	218	23.7%
HSValidate	69	55	22.2%
Total Time	528	375	28.9%
HSph@SF at 1-TB Scale Factor	6.72	9.45	

Table 5 lists detailed response times for each benchmark phase.

Test Result 2: 1500 Versus 9000 Maximum Transmission Unit
One the most commonly tuned parameters is the MTU, which defines the largest packet size that an interface can transmit without the need to fragment the packet. IP packets larger than the MTU require IP fragmentation.

The use of jumbo frames (an MTU value of 9000) improves performance because jumbo frames reduce the number of individual frames that must be sent for a given amount of data, and they reduce the need to separate data blocks into multiple Ethernet frames. They also reduce host and storage CPU utilization.

Figure 4 shows the performance improvement with a larger MTU (9000).

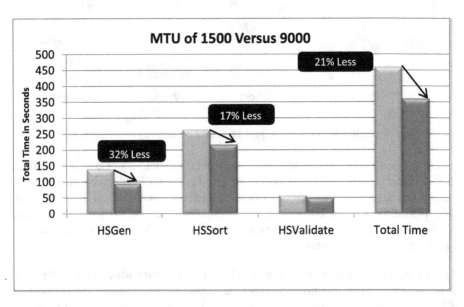

Fig. 4. MTU of 1500 Versus **9000**

Table 6 lists detailed response times for each benchmark phase.

Table 6. MTU of 1500 versus 9000

Phase	Bonding (Multiple NICs at 1500 MTU)	Bonding (Multiple NICs at 9000 MTU)	Percentage improvement
HSGen	140	95	32.1%
HSSort	264	217	17.8%
HSValidate	56	49	12.5%
Total Time	460	361	21.5%
HSph@SF at 1-TB Scale Factor	7.71	9.81	

Test Result 3: Two-vNIC Bonding Versus Three-vNIC Bonding

Cisco UCS VIC technology supports up to 256 virtual NICs (vNICs). Tests with three vNICs provided slight performance improvement, as shown in Fig. 5.

[[PLS CHANGE THE CALLOUTS AS FOLLOWS:]]
Two-vNIC Bonding Versus Three-vNIC Bonding
Time in Seconds
(2 NICs)
(Multiple NICs)

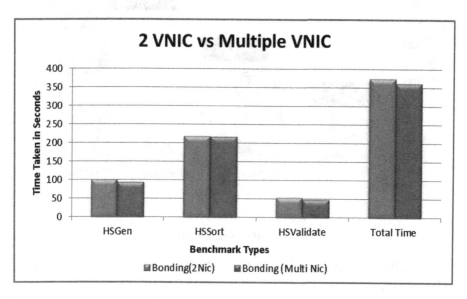

Fig. 5. Two-vNIC Bonding versus Three-vNIC Bonding

Table 7 lists detailed response times for each benchmark phase.

Storage Tuning

Optimal configuration of the storage system is extremely important to achieve the best application performance. In most cases, servers with internal direct-attached storage (DAS) provide the best performance and price-to-performance ratios. Two popular storage controller options are RAID controllers and host bus adapters (HBAs). In

Table 7. Two-vNIC Bonding versus Three-vNIC Bonding

Phase	Bonding (2 NICs)	Bonding (Multiple NICs)	Percentage improvement
HSGen	102	95	6.86%
HSSort	218	217	0.45%
HSValidate	55	49	10.9
Total Time	375	361	3.73%
HSph@SF at 1-TB Scale Factor	9.45	9.81	

addition to RAID functions, RAID controllers offer advanced self-monitoring, analysis, and reporting technology (SMART) features and write-back or flash-based write cache. SMART features detect and report the health of the disk drives beyond the capabilities of JBOD. Caching can improve data load performance in Hadoop deployments. This section describes best practices based on the tests conducted on the Cisco UCS Integrated Infrastructure for Big Data cluster.

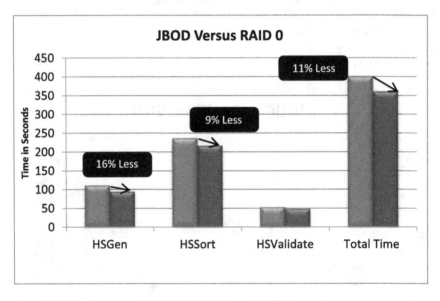

Fig. 6. JBOD Versus RAID 0

Table 8 lists optimal settings for the Cisco 12-Gbps SAS modular RAID controller for Hadoop deployments.

Test Result 4: JBOD Versus RAID
JBOD and RAID 0 work similarly. The main difference pertaining to performance is the effect of controller caching. Figure 6 shows better performance with RAID than with JBOD. The controller cache (a 2-GB module was used in these tests) optimizes writeback operations when the workload is based on large sequential read and write processing.

Table 9 lists detailed response times for each benchmark phase.

Table 8. Optimal RAID controller settings for Hadoop

Parameter	Setting
RAID	RAID 0 for individual disk drives
Controller Cache	Always write back NoCacheBadBBU Read ahead
Stripe Size	1024 KB
Disk Drive Cache	Enabled (read) (Cisco firmware does not allow the write cache to be enabled on disk drives.)

Table 9. JBOD versus RAID 0

Phase	JBOD	RAID 0	Percentage improvement
HSGen	111	95	16.84%
HSSort	237	217	9.22%
HSValidate	53	49	8.16%
Total Time	401	361	11.08%
HSph@SF at 1-TB Scale Factor	8.82	9.81	

Operating System Tuning

Changing some system settings in Linux can increase overall performance. This section discusses these changes and their benefits. Table 10 lists some of the OS performance settings best for Hadoop.

Table 10. Operating system settings

Parameter	Value
vm.dirty_background_ratio	1
vm.swappiness	0
vm.overcommit_memory	0
net.core.rmem_max	16777216
net.core.wmem_max	16777216
net.core.netdev_max_backlog	10000

In addition, the following settings for /etc./security/limits.conf are recommended:

- root soft nofile 64000
- root hard nofile 64000
- hadoop soft nproc 32768
- hadoop hard nproc 32768
- hadoop soft nofile 32768
- hadoop hard nofile 32768

File System Tuning

Different Linux distributions use different default file systems. Testing has shown that XFS seems to be better than Ext3 or Ext4 for Hadoop. XFS is a high-performance journaling file system that was initially created by Silicon Graphics for the IRIX operating system and later ported to Linux. XFS has a large number of features that make it suitable for deployment in an enterprise-level computing environment that requires implementation of very large file systems.

XFS has very bad performance out of the box. Unlike with Ext4, the file system needs to be formatted with the right parameters to perform well. And if you don't specify the parameters correctly, you need to reformat the file system because you can't change the parameters later. The main parameter that the tests reported here found useful to tune is **agcount**: the number of allocation groups. Allocation groups enable simultaneous I/O processing by multiple application threads. XFS splits the file system into multiple allocation groups to help increase parallelism, because each allocation group has its own set of locks. It is important to create as many allocation groups as you have hardware threads. If the server has a dual CPU configuration with 16 cores and 32 threads with hyperthreading, an **agcount** value of 32 is recommended for best I/O performance.

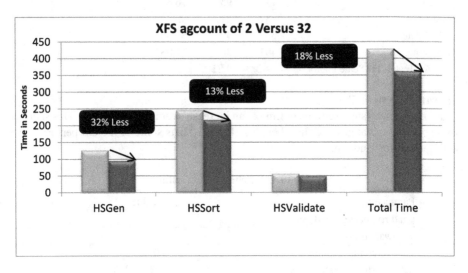

Fig. 7. XFS agcount of 2 Versus 32

XFS supports several mount options that can influence behavior. XFS allocates inodes according to their on-disk locations by default. However, because some 32-bit user-space applications are not compatible with inode numbers greater than 2^{32}, XFS allocates all inodes in disk locations that result in 32-bit inode numbers. This behavior can lead to decreased performance on very large file systems (systems larger than 2 terabytes [TB]), because inodes are skewed toward the beginning of the block device, and data is skewed toward the end. To address this scenario, the **inode64** mount option is recommended.

Linux records information about the time when files were created, last modified, and last accessed. There is a cost associated with recording the last access time. The **noatime** attribute tells the file system not to record the last-accessed time for the file and is recommended for Hadoop deployments.

Test Result 5: XFS with agcount of 2 Versus 32

Tests for conducted with allocation groups of 2 and 32. As shown in Fig. 7, an optimal allocation count is critical for optimizing XFS for Hadoop.

Table 11 lists detailed response times for each benchmark phase.

Table 11. XFS agcount of 2 versus 32

Phase	Agcount = 2	Agcount = 32	Percentage improvement
HSGen	126	95	32.63%
HSSort	246	217	13.36%
HSValidate	56	49	14.29%
Total Time	428	361	18.56%
HSph@SF at 1-TB Scale Factor	8.27	9.81	

Another important OS setting is the CPU frequency and scaling governor (Table 12). The performance mode is recommended for high-performance Hadoop deployments.

Table 12. CPU Governor options in Linux

Governor	Description
ondemand	Dynamically switch between CPUs available if 95% CPU load is reached.
performance	Run the CPU at maximum frequency. This mode is recommended for high-performance Hadoop deployments.
conservative	Dynamically switch between CPUs available if 75% CPU load is reached.
powersave	Run the CPU at the minimum frequency.
userspace	Run the CPU at user-specified frequency.

Transparent huge pages is a commonly used option that works well in most instances, including with Hadoop. However, a problem arises with one feature of transparent huge pages called compaction. This feature defragments memory at the cost of CPU cycles. Testing has shown better performance with compaction disabled. This option can be set with the following command:

```
#          echo          never          >
/sys/kernel/mm/redhat_transparent_hugepages/defrag
```

Linux swappiness is a kernel process that finds memory content that has not been used in a while and copies it to the hard drive. The swappiness value can be adjusted from 0 to 100. In most versions of Linux, the default value is 60. The tests reported here show that turning off swappiness (setting swappiness to 0) is optimal for Hadoop deployments. This option can be set with the following command:

```
sysctl -w vm.swappiness=0
```

The I/O scheduler is another important performance tuning option. The recommended I/O scheduler setting for Hadoop is Completely Fair Queuing (CFQ). CFQ is the default setting in some Linux distributions, and it can increase performance by 2 or 3 percent. This option can be set with the following command:

```
echo cfq > /sys/block/sd*/queue/scheduler
```

Hadoop Tuning

Out of the box, many Hadoop settings are not optimized for best performance. HDFS provides storage for all the data and is a core component of Hadoop. Fine-tuning the settings here can produce significant performance improvements. The settings discussed in this section have been tested and will provide improved speed for heavy workloads.

The Hadoop block size defines the number of input splits for a file. Each input split is replicated three times (by default) across the cluster. Map tasks typically operate on these input splits. The number of input splits determines the number of map tasks.

The total read time on hard disk drives consists of seek time (finding the first block of the file) and transfer time (the time needed to read contiguous blocks of data). When dealing with hundreds of terabytes or petabytes of data, these times become significant. Hadoop handles this processing by having lots of map tasks reading and writing data in parallel. However, processing can benefit by limiting the number of tasks running in parallel, because having too many map tasks trying to read and write data is inefficient. The best approach is a balanced number of input splits and map jobs, because having too few map jobs also reduces performance, just as does having too many.

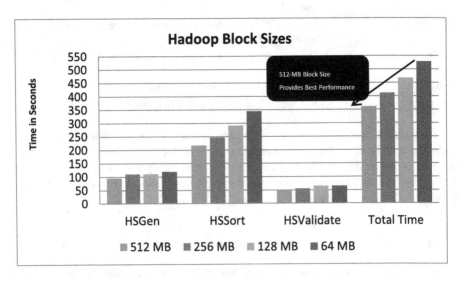

Fig. 8. Impact of Block Sizes

The recommended balance uses this calculation:

Number of launched map tasks = Total size/Input split size (or block size)

Using this formula, for a 1-TB data set with a 64-MB block size, Hadoop would run 15,120 map tasks; with a 512-MB block size, it would run 2160 map tasks.

Test Result 6: HDFS Block Sizes

Tests were conducted with block sizes of 64, 128, 256, and 512 MB. As shown in Fig. 8, 512 MB provided the best performance for the TPCx-HS benchmark. Additional tests conducted with customer workloads reached the same conclusion: that for MapReduce-based applications, larger block sizes provide the best performance.

The configuration is set in hdfs-site.xml as shown in Table 13.

Table 13. hdfs-site.xml Settings

Parameter	Value
dfs.blocksize	512 MB
dfs.datanode.drop.cache.behind.writes	True
dfs.datanode.sync.behind.writes	True
dfs.datanode.drop.cache.behind.reads	True

The general rule for memory tuning is to use as much memory as you can without triggering swapping. The parameter **mapred.*.child.java.opts** can be used to set the task memory. The recommended heap size for both map and reduce tasks is 2 GB, and **ulimit** was set to 4 GB (double the heap size used by all JVM processes) for this workload.

Another important tuning option is to reduce the map disk spill. Mappers generate intermediate data output, which is stored in a memory buffer that is determined by the **io.sort.mb** parameter. This chunk of memory is part of the map JVM heap space. As soon as the threshold is reached (**io.sort.spill.percent**), the content is written to the local disk. This content is called spill. To store the record, the Hadoop framework uses the **io.sort.record.percent** value of the memory allocated by **io.sort.mb**. Performance problems occur when you spill records to disk multiple times. The values of the map output records counter and spilled record counters can be checked for each job, and you can allocate the appropriate memory buffer and the **io.sort.spill.percent** value to use nearly full capacity to enhance Hadoop job performance. These are the recommended settings:

```
io.sort.mb = 1024 MB
```

```
io.sort.spill.percent = .98
```

The number of mappers and reducers is critical to get the best performance. This configuration is based on a 16-node cluster, with one server configured as the name node and 15 servers configured as data nodes, and each server with two CPUs with a total of 48 threads. A slight oversubscription of the number of mappers and reducers to the number of cores should be used, because reducers typically don't start at the same time as mappers. Given the 48 threads in the system under test, allocate 36 threads for mappers and 30 threads for reducers for each node. (This number will vary based on the scale factor of the workload and the system configuration.) The number of HDFS blocks in the input files usually determines the number of mappers. The tuning goal of mappers should be to control the number of mappers and the size of the job. When dealing with large files, Hadoop splits the file into smaller chunks so that the mapper can run it in parallel. However, initializing the new mapper job usually takes a few seconds, creating overhead that should be reduced. To determine the optimal number, several iterations were run.. The configuration for **mapred-site.xml** is shown in Table 14.

Table 14. mapred-site.xml Settings

Parameter	Value
Mapred.map.tasks	540
Mapred.reduce.tasks	450
mapred.tasktracker.map.tasks.maximum	36
mapred.tasktracker.reduce.tasks.maximum	30
mapred.map.child.java.opts	-Xmx800 m -Xms800 m -Xmn256 m
mapred.reduce.child.java.opts	-Xmx1200 m -Xmn256 m
mapred.child.ulimit	4096 MB
io.sort.mb	1024 MB
io.sort.factor	64
io.sort.record.percent	0.15
Io.sort.spill.percent	0.98
mapred.job.reuse.jvm.num.tasks	−1
mapred.reduce.parallel.copies	20
mapred.reduce.slowstart.completed.maps	0
tasktracker.http.threads	120
mapred.job.reduce.input.buffer.percent	0.7
mapreduce.reduce.shuffle.maxfetchfailures	10
mapred.job.shuffle.input.buffer.percent	0.75
mapred.job.shuffle.merge.percent	0.95
mapred.inmem.merge.threshold	0
mapreduce.ifile.readahead.bytes	16777216
mapred.map.tasks.speculative.execution	False

Also, in the hdfs-site.xml file, the **io.sort.factor** parameter controls the number of concurrent streams from the map output that are merged and saved to disk. For heavy workloads with many map tasks, this value should be increased from 10 to 64, to increase the number of streams merged at the same time. This setting has been tested and shown to increase performance, but it should be used with caution on other equipment because it could lead to instability by overworking the system.

Under heavy workloads, Hadoop can launch many thousands of jobs, each of which runs for only a short period of time, and each launching a separate JVM. By default, each JVM must be started and torn down every time. Obviously, this approach is inefficient. It can be improved by changing the parameter **mapred.job.reuse.num.tasks** in the mapred-site.xml file. Change this parameter to −1, and JVMs can be reused for an unlimited number of jobs. This change also helps the platform take full advantage of Java's just-in-time (JIT) compilation, because the JVM does not need to be compiled each time.

Compression can significantly improve Hadoop performance by reducing disk I/O processing and network traffic. It also reduces the amount of disk space used. The TPCx-HS requirements enforce the use of uncompressed job output, but intermediate map output compression is allowed. Table 15 lists the recommended compression parameters.

Table 15. Compression Parameters

Parameter	Value	Description
mapred.output.compress	False	Compression allowed for the MapReduce output
mapred.compress.map.output	True	Compression allowed for intermediate map output
mapred.map.output.compression. codec	org.apache.hadoop.io. compress.SnappyCodec	

Another important tuning parameter is file buffer size, a setting in core-site.xml. The recommended setting for the **io.file.buffer.size** parameter is 131072.

Test Result 7: End-to-End I/O and Network Utilization
Sort workloads are popular in the Hadoop space. TPCx-HS enables fair comparisons to be made between software and hardware systems. It also exercises various subsystems. Figure 9 shows disk read, disk write, network read, and network write utilization from one of the nodes for an end-to-end run.

[[PLS CHANGE CALLOUTS AS FOLLOWS:]]
…Resource Utilization Across Various Phases of Job Processing
Peak Write Throughput Is 2.81 GBps
…Is 1.74 GBps
…Peak Write Throughput Is 2.51 GBps
HSValidate Phase
HSSort Phase
Network I/O Send Peak Throughput Is 1.65 GBps, and Receive Peak Throughput Is 1.92 GBps
Network I/O Send Peak Throughput Is 1.65 GBps, and Receive Peak Throughput Is 1.68 GBps
Network I/O Receive
Network I/O Send

As shown in Fig. 9, in the HSGen phase, peak write throughput is 2.81 GBps, which means that each drive is performing at 117 GBps. This equates to 2.81 × 15 = 42 GBps write throughput per cluster. During the shuffle phase, aggregate read bandwidth is 26 GBps, and during the reduce phase, aggregate write bandwidth is 38 GBps. The peak network bandwidth utilization was 1.8 GBps: about 75 percent of dual 10 Gbps connectivity.

Test Result 8: End-to-End CPU Utilization
One frequently asked question relates to CPU utilization. Figure 10 shows the CPU utilization for an end-to-end TPCx-HS run. As noted, CPU utilization was about 97 percent peak at the shuffle and sort phase.

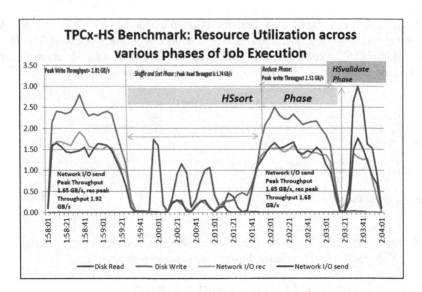

Fig. 9. Resource Utilization across various Job Processing phases

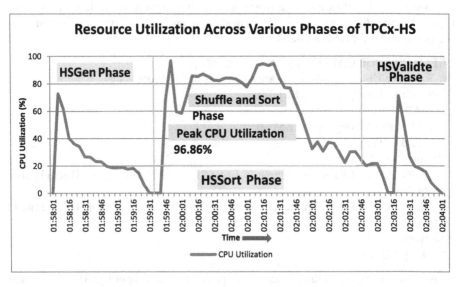

Fig. 10. CPU Utilization Across Various Phases

As observed in the results from tests 8 and 9, the TPCx-HS benchmark exercises the upper boundaries of I/O, network, and CPU processing with Hadoop. This feature makes TPCx-HS a good benchmark standard that enables fair comparison of Hadoop systems, and it also provides a good workload for stress-testing various technologies under development.

6   Conclusion

This document provides a summary of lessons learned from performance tuning for the TPCx-HS benchmark. The tuning parameters and test results have broad applicability across Hadoop-based applications. The test results also address some of the most frequently asked questions about Hadoop system tuning.

References

1. IDC Worldwide Big Data Technology and Services Forecast (2015)
2. Nambiar, R., Poess, M., Dey, A., Cao, P., Magdon-Ismail, T., Da Ren, Q., Bond, A.: Introducing TPCx-HS: the first industry standard for benchmarking big data systems. In: Nambiar, R., Poess, M. (eds.) TPCTC 2014. LNCS, vol. 8904, pp. 1–12. Springer, Cham (2015). doi: 10.1007/978-3-319-15350-6_1
3. Nambiar, R.: A standard for benchmarking big data systems. In: IEEE Big Data Conference, pp. 18–20 (2014)
4. TPCx-HS specification. http://www.tpc.org/tpcx-hs/

Work-Energy Profiles: General Approach and In-Memory Database Application

Annett Ungethüm[✉], Thomas Kissinger, Dirk Habich, and Wolfgang Lehner

Database Systems Group, Technische Universität Dresden,
01062 Dresden, Germany
{annett.ungethum,thomas.kissinger,
dirk.habich,wolfgang.lehner}@tu-dresden.de

Abstract. Recent energy-related hardware developments trend towards offering more and more configuration opportunities for the software to control its own energy consumption. Existing research so far mainly focused on finding the most energy-efficient hardware configuration for specific operators or entire queries in the database domain. However, the configuration opportunities influence the energy consumption as well as the processing performance. Thus, treating energy efficiency and performance as independent optimization goals offers a lot of drawbacks. To overcome these drawbacks, we introduce a model based approach in this paper which enables us to select a hardware configuration offering the best energy efficiency for a requested performance. Our model is a work-energy-profile being a set of useful work done during a fixed time span and the required energy for this work for all possible hardware configurations. The models are determined using a well-defined benchmark concept. Moreover, we apply our approach on in-memory databases and present the work-energy profiles for a heterogeneous multiprocessor.

Keywords: Energy efficiency · In-memory database systems · Benchmarking · Profiles

1 Introduction

Energy consumption has become a crucial factor in data centers [5,7] and already is the limiting factor for the scalability of many-core processors. This limitation can be observed on today's processors, which are not designed to run at their peak performance for a long time because thermal and power related limitations lead to a reduced performance or even dark silicon effects [2]. Driven by the recent advances in the mobile devices sector, technology providing a fine-grained control over the energy consumption of individual cores made its way into the server CPU market. That means, hardware and operating systems offer several control knobs for reducing the energy consumption of a system, e.g., frequency scaling (DVFS) and sleep states. To exploit their full potential, software has to make appropriate use of these control knobs. This exploitation becomes more

© Springer International Publishing AG 2017
R. Nambiar and M. Poess (Eds.): TPCTC 2016, LNCS 10080, pp. 142–158, 2017.
DOI: 10.1007/978-3-319-54334-5_10

complex by introducing hardware heterogeneity as done e.g., by the ARM®
big.LITTLE™ hardware technology [6].

Using these control knobs, the hardware can be configured in a regulated
way influencing the energy consumption as well as the performance. From a
database perspective, we can state that an energy-efficient in-memory database
system must be fast enough to process queries with a certain maximum latency.
However, a low latency query execution does not necessarily correspond to a
low energy consumption [6]. Thus, treating energy efficiency and performance
as independent optimization goals (1) makes the system waste energy because
it uses a high performing but much more power draining configuration or (2)
increases the latency to a point where the query queue grows faster than it can
be processed while operating in an energy-efficient mode. To overcome this issue,
our objective is to avoid both of these traps by applying a combined approach.

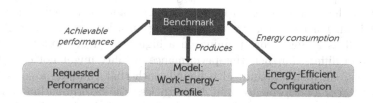

Fig. 1. Model-based approach using a benchmark concept.

That means, instead of choosing a hardware configuration using the control
knobs which always offer the best performance *or* the best energy efficiency, we
have to determine a hardware configuration offering the best energy efficiency
for a *requested or desired performance requirement range*. For selecting this configuration, a model is required providing the most energy-efficient configuration
for a requested performance as depicted in Fig. 1. To enable that, we propose our
solution called *work-energy profiles*. A *work-energy profile* is a set of the useful
work done during a fixed time span and the required energy for this work for
all possible hardware configurations. Based on these *work-energy profiles*, we are
able to select an energy-efficient hardware configuration for a requested performance range. To create such *work-energy profiles* for a concrete hardware and
specific application, we introduce our developed benchmark concept.

Our Contribution and Outline: We begin our presentation with a solution
overview by describing our running example hardware for this paper and introducing our *work-energy profiles* in Sect. 2. In particular, our example heterogeneous multiprocessor system, the ARM® big.LITTLE™ based ODROID-XU3,
offers a rich set of configuration opportunities. Then, we propose our general
benchmark concept for the creation of *work-energy profiles* in Sect. 3. In detail,
we introduce our metrics for evaluating the energy awareness and performance
of software in general. Moreover, we define the overall benchmark sequence and
present several dependencies which have to be considered. Afterwards, we apply

our benchmark concept to a concrete scenario in Sect. 4. Here, we investigate in-memory database systems on our heterogeneous hardware. Using our benchmark, we clearly show a dependency between energy efficiency and typical database main memory access patterns as well as a high variance in performance and energy consumption for different hardware configurations. Based on this investigation, we draw some further conclusions for energy-efficient in-memory database systems in Sect. 5. We conclude the paper with related work in Sect. 6 and a short summary including an outlook in Sect. 7.

2   Solution Overview: Work-Energy Profiles

As already mentioned in the previous section, modern hardware and operating systems offer several control knobs to adjust hardware settings and accordingly influence the performance and the energy consumption. However, a mapping between the hardware configuration, performance and energy efficiency is not always trivial. For example, two cores running on a low frequency might perform as well as a single core running on a higher frequency but their energy consumption differs. Further, this performance equality might not exist for all applications, e.g. if it is bandwidth bound, such that enabling a second core hardly produces a performance gain. Nevertheless, the determination of a hardware configuration offering the best energy efficiency for a desired performance is important for applications.

To capture all hardware configuration possibilities and to consider all hardly predictable effects, we propose to solve this challenge using our so called *work-energy profiles*. A *work-energy profile* is a set of the useful work done during a fixed time span and the required energy for this work for all possible hardware configurations. The *work-energy profiles* have to be determined for an specific application and on a concrete hardware system. Based on these *work-energy profiles*, we are able to select an energy-efficient hardware configuration for a requested application performance range. While this section introduces our *work-energy profiles* as a general solution, we also present our underlying test hardware which is used through the paper as a running example.

2.1   Heterogeneous Test Hardware

In the last years, hardware has been shifted from single CPU to multiprocessors with increasing main memory capacities. At the moment, the hardware is changing again from homogeneous towards heterogeneous systems, mainly to reduce energy consumption to avoid Dark Silicon effects or to increase the system's performance since homogeneous multiprocessors reached several physical limits in scaling [2]. This heterogeneity and the corresponding control knobs to adjust the hardware settings have a non-trivial influence on the performance and the energy consumption. Therefore, we have chosen the ODROID-XU3 as our running example hardware, which is based on the ARM® big.LITTLE™ technology. In this architecture, relatively battery-saving and slower processor cores

(LITTLE) are coupled with relatively powerful and power-hungry ones (big). The ODROID-XU3is operating in the heterogeneous multi-processing mode (HMP) allowing us to freely assign threads to specific big and LITTLE cores. Contrary to earlier ARM® big.LITTLE™ processors, both clusters (LITTE and big) of the ODROID-XU3 can be active at the same time. The 2 GB of main memory are shared for all cores. Table 1 gives an overview of the available core clusters and the respective configuration options.

Table 1. Configuration options of the ARM® big.LITTLE™ ODROID-XU3

	LITTLE Core Cluster	big Core Cluster
Core description	ARM-CortexA7	ARM-CortexA15
Number of Cores	4	4
Frequency range	0.2 GHz–1.4 GHz	0.2 GHz–2.0 GHz
Frequency step range	100 MHz	
Number of freq. steps	13	19
Pipelines	1	3
Features	-	out-of-order execution

Additionally, our hardware system is equipped with on-board power sensors allowing us to measure the power level of individual core clusters and the main memory separately. The system draws around 2.5 W (CPU: 1.1 W) in idle mode. Under load, it draws 10.7 W of which 7.0 W are the power consumption of the CPU. Hence, around two thirds of the drawn power under load is consumed by the CPUs. There is no HyperThreading and only the performance governor (a power policy for the CPU) is available, but the system allows to manually set one of the 13 different frequencies for the LITTLE A7 cluster as well as one of the 19 for the big A15 cluster. Different frequency settings on a completely idle cluster cannot be summarized because a changing frequency on the idle cluster influences the performance on the active cluster for many use-cases (see Sect. 5). This already extends to $19 \cdot 13 \cdot 24 = 5,928$ possible hardware configurations. For demonstration purposes, we refrain from using any specialized instruction sets, e.g. NEON, which increases the number of configurations even further.

2.2   Example Work-Energy Profile

With 5,928 possible hardware configurations for our example hardware, we have a large space of opportunities to satisfy a desired performance range of an application task. The challenge is now to determine the most energy-efficient hardware configuration for a requested performance range. To tackle this challenge, we propose our *work-energy profiles*. To get a deeper understand of our *work-energy profile* approach, Fig. 2 illustrates an example for an application task running on our test hardware. The left chart in this figure shows the corresponding *work-energy profile*. While the performance is plotted on the x-axis, the y-axis shows the energy efficiency. Each dot in this chart represents a specific hardware configuration. As we can see, different hardware configurations offer the same performance range with a high variance in the energy efficiency. Therefore, a work-energy profile depicts the energy efficiency as a function of performance. From this profile, we are able to derive various insights as highlighted in the remainder of the paper. To determine these dots, we executed the application task with each possible hardware configuration and measured the performance as well as the energy efficiency. For the systematic construction of such *work-energy profiles*, we developed a benchmark concept as described in the next section.

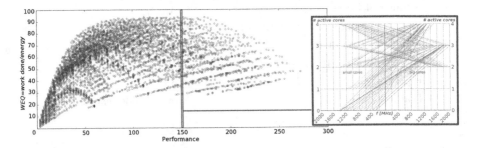

Fig. 2. Work-Energy Profile and a close-up of the highlighted performance range.

Generally, we are able to utilize such profiles to directly identify the most energy-efficient configuration (high energy efficiency value) for a desired performance range and application task. In Fig. 2, we highlighted a specific performance range using a vertical slice. This performance range can be realized with various hardware configurations as depicted in the right chart of this figure. In this chart, the most energy-efficient configuration is highlighted by a thick green line. The x-axis indicates the frequency of the clusters, the y-axis shows the number of active cores. The left side shows the A7 cluster, the right side the A15 cluster. A line connects the configuration of both clusters and forms the complete configuration. The least energy-efficient ones are marked with a thin red line. These close-ups show the variety inside the configurations, which produce the same performance but a different energy efficiency and the most energy-efficient configuration is not necessarily the most obvious one. In Sects. 4 and 5, we apply our approach to an in-memory database system and draw some conclusions regarding energy-efficient database systems.

3   Creation of *Work-Energy Profiles* - Benchmarking

One of our main challenges is the creation of *work-energy profiles* covering a large number of possible hardware configurations. To tackle this challenge, we developed a benchmark concept to examine the behavior of performance and energy efficiency for different hardware configurations in a uniform way. Fundamentally, the same test-case or application task has to be repeated and recorded for all possible configurations on the target systems. Moreover, not only the task but also the test data has to be the same in order to produce comparable results. Therefore, we separated the generation of test-cases and data from the control-flow of our benchmark concept. Generally, an overview of our benchmark concept is depicted in Fig. 3. A concrete implementation and application is presented in Sect. 4, while this section describes the general benchmark concept.

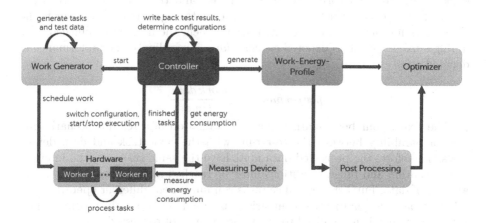

Fig. 3. An overview of the benchmark setup.

As shown in Fig. 3, the `Controller` is the centerpiece of our benchmark. It starts the `Work Generator` which produces tasks and test data. This `Work Generator` has to be adjusted for each application scenario. After the `Work Generator` has finished, the `Controller` chooses the first hardware configuration and starts the first test run. Within a test, the corresponding tasks are processed by every worker, whereas a worker is a thread running on a (virtual) core. The workers count their finished tasks. After a fixed time span, the `Controller` shuts down the threads and collects the number of finished tasks which are later used for calculating the performance. During an active test, the values necessary for the energy computation are recorded by a `Measuring Device`. Depending on the abilities of the `Measuring Device`, the energy computation is either done by the device itself or by the controller. In both cases the final values are collected by the `Controller`. For eliminating odd side effects, a test can be run multiple times. Then, the test runs for the remaining configurations but the same tasks and test data are successively executed. After all configurations have been processed, the `Controller` generates a *Work-Energy-Profile* for the selected task and data as depicted in Fig. 2. This profile can then be used for in-depth analysis and optimization purposes, e.g. for choosing an energy-efficient configuration satisfying the requested performance constraints or for optimizing the applied algorithm.

3.1 Metrics

In addition to this general benchmark sequence as presented above, it is important, which metrics have to be measured at all. In our case, performance and energy efficiency are the relevant metrics and specified in detail as next:

Work and Performance: The *hardware* processes the *work* generated by the Work generator. This *work* consists of a task and the data to be processed. The task is repeated over the same amount of data, e.g. scan of records or hashing of keys for each hardware configuration. Then, the number of finished task during a fixed runtime is denoted as *work done*. Accordingly, the performance is denoted as:

$$performance = \frac{work\,done}{time}$$

The goal of our benchmark is not the evaluation of a real-life scenario but the comparability between the test runs with the same task and data definitions. Therefore, the processed data must have the same size and type in every task execution and break conditions must be reached after processing the same amount of data. Furthermore, since tasks can implement different operations on different data, a quantitative comparison between them is only possible when the performance is normalized to the same amount of processed data.

Energy: The electric power P is the product of the amperage $i(t_1)$ and the voltage $v(t_1)$ measured at t_1 (Eq. 1). Thus, it describes the power consumption of a measured system at a discrete point in time. In contrast, the electrical energy E is the integral over time of the whole power curve consisting of all power values taken during the measurement (Eq. 2). Thus, the consumed energy grows while time passes whereas the power can rise and drop.

The Measuring Device is responsible for determining amperage and voltage. The on-board power sensors of our ARM® big.LITTLE™ hardware satisfy this property. Intel® introduced also on-board energy sensors—called "Running Average Power Limit" (RAPL)—with their Sandy Bridge microarchitecture [4]. Therefore, our benchmark concept can also be applied to Intel® multiprocessors without special instrumentation as long as the corresponding sensors are available. The computation of the energy from these measured values can either be done by the Measuring Device or by the Controller.

$$P(t) = v(t) \cdot i(t) \tag{1}$$

$$E = \int_{t_0}^{t_{\text{end}}} P(t)\,\mathrm{d}t = \int_{t_0}^{t_{\text{end}}} v(t) \cdot i(t)\,\mathrm{d}t \tag{2}$$

An optimization for power is necessary, e.g., for thermal chip design or for dimensioning the necessary cooling, but the reduction of the electricity bill and the extension of the life span of a battery charge require energy optimization, i.e. the costs on the electricity bill are calculated from the energy drawn since the last meter reading. Hence, the primary goal for increasing energy efficiency is the reduction of the overall energy drawn by the system while doing the same amount of work. Only reducing the power, e.g. by reducing the core frequencies, could lead to longer execution times and therefore to a higher energy consumption.

Energy Efficiency: To achieve the objective of reducing the overall energy consumed by the system for a specific amount of work, the natural decision for quantifying the energy efficiency is to calculate it as the quotient of *work done* and *consumed energy* [5,7,9]. Accordingly, we call this relation the *Work-Energy Quotient (WEQ)*.

$$WEQ = \frac{work\,done}{energy}$$

Since $work\,done/time$ equals the $performance$ for a certain operation, the mentioned works rewrite the quotient as follows:

$$\frac{work\,done}{energy} = \frac{work\,done}{power \cdot time} = \frac{performance}{power}$$

As already discussed, the power level can change during the measurement and the energy is its integral over time. Thus, the real power values cannot be restored from the energy, although the rewritten equation implies that this was possible. Vice versa, energy cannot be computed by an average power value when the required time is not necessarily the time needed for the computation but a part of a potentially normalized performance value. For this reason and for avoiding a confusion of execution time and normalized performance values, we argue to use $work\,done/energy$ instead of $performance/power$. For not mistaking the WEQ for energy efficiency definitions in other fields, e.g. the energy conversion efficiency, we do not just call this definition *energy efficiency (EE)* but *work-energy quotient*.

3.2   Benchmark Setup and Dependencies

Up to now, we described our general benchmark sequence and defined our measured metrics. A full benchmark tests the same tasks and data configuration on all possible *hardware configurations* and these hardware configurations have to be set by the `Controller`. A *hardware configuration* contains the settings of hardware components which can be adjusted. Thus, it depends on the hardware system the benchmark is running on. A common configuration could consist of the following parts:

- Bitmask for defining active workers (w_i)
- Frequency of the physical cores (f_i)
- If available: Bitmask indicating if a specialized instruction set is used, e.g. SSE (is_i)

A configuration containing these options could be described by the vector $\{w_1, .., w_n, f_1, .., f_m(, is_1, .., is_m)\}$ with $n = m * hyperthreads\ per\ core$. Such a description is used to iterate over all possible hardware configurations by our `Controller`. For our running example hardware, the ODROID-XU3, a configuration consists of an 8-bitmask $\{w_0, ..., w_7\}$ and two frequencies $\{f_{A7}, f_{A15}\}$. The bitmask indicates which cores are actively processing tasks. Since there are no hyperthreads, one bit per core is sufficient. The frequencies can be adjusted per cluster. Hence, there are two frequencies in every configuration, one for the A15-cluster and one for the A7-cluster.

Furthermore, we have to consider the following aspect: The generated work, the processing speed of the hardware and the specifications of the measuring device influence the accuracy of the tests. This implies that certain aspects have to be considered before implementing this benchmark. First, the `Measuring Device` must be suitable for the power range of the system, e.g. an accuracy of 1 W might be accurate enough for a system drawing between 100 and 500 W but not for one drawing between 1 and 10 W. In our implementation, we use the integrated current sensors of the ODROID-XU3 which fulfill this requirement. Second, the runtime of a single test must be long enough to gain significant results. In detail it has to fulfill the following requirements:

- Compensate for out-dated values due to the update frequency of the measurement device. Ideally a new test always starts right after an update cycle of the measurement device.
- Compensate for varying power values, e.g. if the power level changes during a test case but only one power value is recorded, the measurement result of this particular test case has only limited expressiveness because the energy is only computed from this single power value.
- Finish a significant amount of *work* on all workers, e.g. if half of the cores are not even able to process one request, there is no work done which can be compared even if some of the cores would have finished much faster than the other ones.

The integrated sensors of our example hardware only update their values approximately every quarter second. Hence, depending on the exact task, we run every test between 5 and 10 s to gain between 20 and 40 values to compute the energy from and to finish some tasks on every worker even in very low performing configurations.

4   Application of *Work-Energy Profiles*

Energy awareness of database systems has emerged as a critical research topic [6,7]. In our research, we focus on energy-efficient in-memory database systems. Therefore, CPU and main memory are the main hardware components of interest. Here, the performance and energy efficiency of a hardware configuration depend on a multitude of factors (e.g., data characteristics and size, operators types, etc.). Moreover, main memory bandwidth and latency are limiting factors that could cause a non trivially predictable hardware behavior. To get a deeper understanding of this issue, we have investigated in-memory databases with our benchmark concept using fine-grained memory access patterns which are highly utilized. Nevertheless, we are also able to investigate complete operators or queries and can create *work-energy profiles* for such database tasks with our approach. Here, the advantage would be that we would have the *work-energy profiles* per operator or query and then we can directly use them. However, there are a lot of disadvantages for such an approach: (i) operators and queries are of complex nature, and (ii) database systems usually execute multiple queries simultaneously, which results in multiple operator types running in parallel. For these reasons, it makes much more sense to explore the configuration space on a fine-grained level of main memory access patterns.

4.1 Typical Database Memory Access Patterns Under Test

As already argued, main memory is the bottleneck for in-memory database systems and therefore, we use four different low-level but basic work operations with significantly different memory access patterns. For each low-level operation, we implemented a corresponding task and data definition for our `Work Generator`:

(1) **Compute-Intensive:** This class simulates work operations that do not involve any main memory utilization, e.g. the solution of mathematical equations. In our implementation, our task includes taking 512 square roots per iteration.

(2) **Scan-Operation:** This workload consists of main memory/cache-bound operations like column scans exhibiting a sequential main memory access pattern including reading and writing operations. The implementation is done using a `memory compare` operation with 128 KB per iteration.

(3) **Lookup-Operation:** This class also represents main memory/cache-bound operations like lookup exhibiting a random main memory access pattern including reading and writing operations. The lookup is done using 32 B in each iteration

(4) **Copy-Operation:** This work class simulates a data copy operation for the creation of intermediate results. As test data, we used 128 KB per iteration.

As described in Sect. 3, the obtained measurements of our benchmark include the performance in iterations/second and the WEQ in iterations/Joule.

4.2 Work-Energy Profiles

Then, we executed our benchmark for each low-level memory access pattern in an isolated way for our selected hardware, to create a specific *work-energy profile* for each of them and to get a first impression of how the test system responds. Figure 4 shows the resulting four *work-energy profiles*. As we can see in this figure, the profiles show a completely different shape. Again, each dot in each diagram represents a hardware configuration and for each hardware configuration, we determined a performance (x-axis) and a WEQ value (y-axis). That means, each low-level operation shows a different behavior with regard to performance and energy efficiency. All profiles have in common, that there are hardware configurations offering the same performance with a different energy efficiency value.

If we look at the compute-intensive operation, this operation has its peak WEQ at $\sim 1/5$ of the maximum performance (see Fig. 4(a)). Afterwards, the performance is further increasing at the cost of the energy efficiency. Furthermore, there are hardware configurations available with a high performance value but less energy efficiency. In contrast to that, the peak WEQs of the other low-level operations are (i) Scan-Operation at $\sim 1/3$, (ii) Lookup-Operation at $\sim 3/4$, and (iii) Copy-operation at $\sim 1/2$ of the maximum performance. Then, the performance increases with decreasing the energy efficiency.

As we can observe in Fig. 4, each low-level operation has its own *work-energy profile* shape. Based on that profiles, we are able to determine a hardware configuration which offers the best energy efficiency for a requested performance range. These hardware configurations are the pareto efficient frontiers of the profiles. At the moment, we can only state that the profile's shapes and the pareto frontiers are different. The question now is, why and to which extent they are different and if there are even similarities recognizable. To answer this question, we have to analyse the profiles in detail as done in the following section.

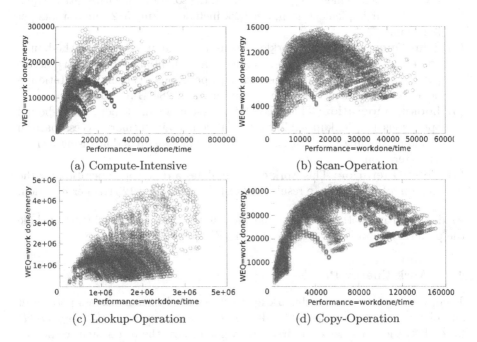

Fig. 4. *Work-energy profiles* for low level memory access patterns.

5   Analysis of Application *Work-Energy Profiles*

For showing why the profiles look the way they do, we exemplarily explain a profile in more detail. We choose the Copy-Operation (memory copy operation) since it includes two main basic tasks a database has to process: reading and writing data, e.g. key-value stores mainly process put- and get-requests where relatively small chunks of data are read from and written to the main memory.

5.1   Intra-Profile Observations

It is useful to know if the same operation has a different profile for different implementations or data sizes. For our memory copy operation, we varied the size of the blocks which are copied. Our original setup copied 32 MB at once. Then, we varied it to copy it in 128 KB and 512 KB blocks (see Fig. 5a). While 128 KB fit well into the L2 cache on both clusters of our example hardware, the 512 KB block leaves no space for any other data in the L2 cache of the A7 cluster. Since the *performance* and peak *WEQ* was growing while we shrinked the block size, we chose a block size of only 32 Bytes for the last implementation, which is only a fraction of the page size of 4 KB. For being able to compare different implementations of the same use-case, the *performance* and *WEQ* need to be normalized to a common measure. Using the performance or WEQ at a certain hardware configuration, e.g. the lowest performing one, still results in different absolute reference measures for differently performing implementations. Therefore, we used the smallest block size of the processed data as a common reference measure. We define *work done = 32 Bytes copied* and calculate the *performance* and *WEQ* accordingly.

The best performing implementation copies the data in blocks that fit into both L2 caches. The *performance* and *WEQ* increase until a peak is reached. Afterwards, the performance can still be increased at the cost of the energy efficiency. As expected, the performance and the energy efficiency decrease when sizes are chosen which do not fit into the cache of the clusters. This effect is especially visible when copying 32 MB blocks. The possibility to gain any performance after the *WEQ* has reached a peak, has massively decreased.

The reason for this behavior is illustrated in Fig. 6. It shows the performance and WEQ for the configurations which contain only one active cluster. In Fig. 6a the frequency on the active cluster is increased, in Fig. 6b the frequency on the idle cluster is increased. In both cases the WEQ does not grow anymore for any core configuration when switching to the higher frequencies. When changing the frequency on the A15-cluster it even decreases after ≈1 GHz. This is most likely because the power consumption grows superlinearly when increasing the frequency. When looking at the number of active cores on the A15 cluster, the energy efficiency decreases when enabling more than two cores. Additionally, the mean performance decreases in Fig. 6b while in Fig. 6a it increases by an insignificant amount.

The worst performing implementation is the 32 B-block implementation. The maximum *performance* and *WEQ* increase until a peak is reached. Then the

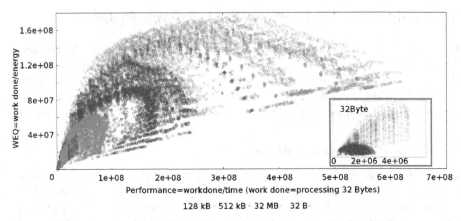

(a) Read and write back of different data sizes, normalized to 32 Byte.

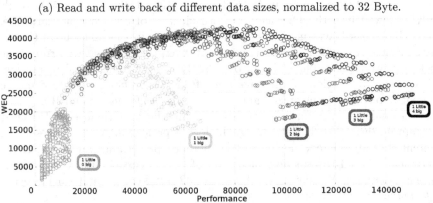

(b) Reduced Work-Energy profile to only show the core configurations producing the optimum for a read/write operation.

Fig. 5. The same operation differently implemented (i.e. varying the block size) produces different profiles. Depending on the implementation they are either memory or bandwidth bound.

WEQ drops while the *performance* hardly increases anymore. There is no configuration which can improve the performance at the expense of the energy efficiency. This behavior is very similar to the lookup test-case (Fig. 4c) where equally small data chunks are accessed. The high variance between different implementations of the same test case on the one hand, and the similarity between different test cases with a similar memory access pattern on the other hand, suggest that the underlying memory access pattern influences the energy and performance behavior of the system much more than the actually performed higher level operation.

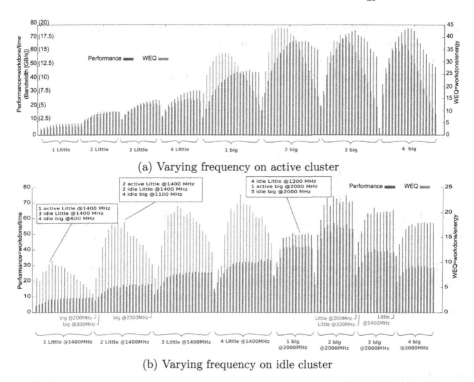

(a) Varying frequency on active cluster

(b) Varying frequency on idle cluster

Fig. 6. A subset of a memory bound use-case to show where the different energy efficiencies for the same performance range come from.

5.2   Inter-Profile Observations

When looking at a single profile, e.g. Fig. 4, there are recurring arcs which differ in their length, height and width. Since they appear, more or less significantly, in all profiles, they are likely not to be a random pattern. Figure 7d shows the core configurations which belong to the most energy-efficient configurations of the profile. They are derived by dividing the profile into 50 equally large performance ranges and selecting the maximum *WEQ* in every of them. A visualization of the configurations which belong to the core configurations in Fig. 5b shows how the arcs in the profiles are generated (for the Copy Operation). Each core configuration spans an arc.

Moreover, Fig. 5b shows that only a few configurations serve the pareto frontier of this profile. Figure 7 shows the optimal core configurations for the test cases from Sect. 4. While they differ for each test case, they are always only a subset of all possibilities. Hence, the pareto frontier is made of a few combinations of active cores.

In our example in Fig. 5b, one configuration, namely four A15 and one A7 core, even covers a wide range of the optimum without ever falling significantly under the optimum of any performance range. Only very small performances

cannot be reached. Hence, by staying at this core configuration and only changing the frequency depending on the requested performance, we always operate
with a good energy efficiency while avoiding task switching and data movement
overhead resulting from a changed CPU set. We call such configurations *robust
configurations*.

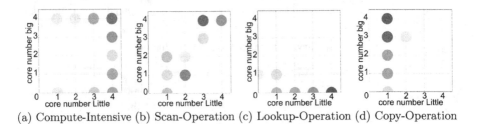

(a) Compute-Intensive (b) Scan-Operation (c) Lookup-Operation (d) Copy-Operation

Fig. 7. Optimal core usage for low level memory access patterns. The value of the color
indicates the frequency of the configuration among the most efficient configurations.
The x-axis shows the active cores on the A7 cluster and the y-axis the active cores on
the A15 cluster.

6   Related Work

Generally, the need for energy efficiency in data centers has already been identified a few years ago [5]. However, before being able to develop a method for
energy savings, the developer is required to know where the energy is spent and
which performance levels are to be expected. An exhaustive investigation on
available hardware has already been made by Tsirogiannis et al. [7]. But this
work did explicitly not focus on energy-efficient hardware configurations and
investigated now outdated hardware with only homogeneous cores.

A naive and more generally applicable approach is simply taking the vendor
specifications of the hardware components. However, these specifications usually
only provide the maximum power consumption [1] and experiments have shown
that they do not reflect the real usage of a system [10]. Hence, a more sophisti-
cated approach has to be found. An analytical model has been implemented by
Xu [10]. But the more complex the hardware is, the more complex this model
gets. Götz et al. present a benchmark-based approach [3], but only apply it for
homogeneous systems.

By applying the TPCH-H queries on an ARM® big.LITTLE™ system,
Mühlbauer et al. have shown that the energy consumption varies heavily depending on the query [6]. Not only the execution plan and the operators themselves
but also the size and type of the data to be processed influence the performance
and the energy consumption. There are countless possible set sizes and different
implementations of query optimizers and operators. Analyzing the workload of
a query on an operator or even query level is too coarse grained to consider all

of these possibilities. It could only produce a model valid for a specific DBMS on a specific hardware and operating-system combination. For this reason we apply a more universal low-level approach considering the hardware component utilization.

7   Conclusion and Future Work

In this paper, we addressed the challenge of benchmarking of *work-energy profiles* to determine the influence of different hardware configurations on the energy efficiency and the performance for a specific application running on a concrete hardware. Aside from presenting *work-energy profiles* in general, we introduced our benchmark concept and applied our approach to an in-memory database on a heterogeneous hardware system. Furthermore, we analyzed the profiles for our application scenario in detail and presented interesting insights. Additionally, by using our *work-energy profiles*, the best possible energy efficiency and the corresponding hardware configuration can be found for a certain required performance range.

Fundamentally, our *work-energy profiles* are the foundation of our vision of an energy-control loop [8]. This energy-control loop addresses the topic of software-controlled hardware reconfigurations at runtime for data management systems. This way, our energy-control loop is able to run the system in an energy-efficient hardware configuration while still being able to maintain certain query latency constraints, especially in times of heavy load. Since the workload is a moving target, the loop is continuously running and adapts the hardware configuration.

Acknowledgments. This work is partly funded by the German Research Foundation in the Collaborative Research Center 912 "Highly Adaptive Energy-Efficient Computing" and within the Cluster of Excellence "Center for Advancing Electronics Dresden" (Orchestration Path).

References

1. ACP - the truth about power consumption starts here. AMD White Paper (2010)
2. Esmaeilzadeh, H., Blem, E., Amant, R.S., Sankaralingam, K., Burger, D.: Dark silicon and the end of multicore scaling. In: ISCA
3. Götz, S., Ilsche, T., Cardoso, J., Spillner, J., Kissinger, T., Aβmann, U., Lehner, W., Nagel, W.E., Schill, A.: Energy-efficient databases using sweet spot frequencies. In: UCC 2014 (2014)
4. Hähnel, M., Döbel, B., Völp, M., Härtig, H.: Measuring energy consumption for short code paths using RAPL. SIGMETRICS Perform. Eval. Rev. **40**(3), 13–17 (2012)
5. Harizopoulos, S., Shah, M., Meza, J., Ranganathan, P.: Energy efficiency: the new holy grail of data management systems research. arXiv preprint arXiv:0909.1784 (2009)
6. Mühlbauer, T., Rödiger, W., Seilbeck, R., Kemper, A., Neumann, T.: Heterogeneity-conscious parallel query execution: getting a better mileage while driving faster! In: DaMoN (2014)

7. Tsirogiannis, D., Harizopoulos, S., Shah, M.A.: Analyzing the energy efficiency of a database server. In: SIGMOD (2010)
8. Ungethüm, A., Kissinger, T., Habich, D., Lehner, W.: Energy elasticity on heterogeneous hardware using adaptive resource reconfiguration live (demo). In: SIGMOD, pp. 2173–2176
9. Wang, J., Feng, L., Xue, W., Song, Z.: A survey on energy-efficient data management. SIGMOD **40**(2) (2011)
10. Xu, Z.: Building a power-aware database management system. In: IDAR (2010)

Erratum to: Performance Evaluation and Benchmarking

Raghunath Nambiar[1](✉) and Meikel Poess[2]

[1] Cisco Systems, Inc., San Jose, CA, USA
rnambiar@cisco.com
[2] Oracle Corporation, Redwood City, CA, USA

Erratum to:
R. Nambiar and M. Poess (Eds.):
Performance Evaluation and Benchmarking, LNCS 10080,
DOI: 10.1007/978-3-319-54334-5

In the original version, there is an error in the title on the cover and the inner title page. It must read "Internet of Things".

The updated original version of this book can be found at DOI: 10.1007/978-3-319-54334-5

© Springer International Publishing AG 2017
R. Nambiar and M. Poess (Eds.): TPCTC 2016, LNCS 10080, p. E1, 2017.
DOI: 10.1007/978-3-319-54334-5_11

Performance and Energy Analysis Using Transactional Workloads

Anastasia Ailamaki[1,2]([⊠]), Danica Porobic[1], and Utku Sirin[1]

[1] EPFL, Lausanne, Switzerland
{anastasia.ailamaki,danica.porobic,utku.sirin}@epfl.ch
[2] RAW Labs, Lausanne, Switzerland

Online Transaction Processing (OLTP) is a multi-billion-dollar industry and one of the most important and demanding database applications. Innovations in OLTP continue to attract significant attention from established industry vendors, startups and a plethora of academic groups worldwide. OLTP applications are characterized by many concurrent requests that typically read about a dozen and write a handful of data items each. The users of the system expect predictably low response times and high availability regardless of the degree of concurrency or the size of data. Thus it is not surprising that a lot of innovations focus on improving the systems to utilize abundant parallelism present in modern multicore servers while ensuring efficient utilization of the microarchitectural resources of each processor core and achieving good energy efficiency.

Analysis of the behavior of existing software systems on modern hardware is the essential precursor to the design of more efficient hardware and software systems of the future. The goal of this process is learning as much as possible about the intrinsic properties of the target workload across different OLTP systems and hardware platforms. Choosing appropriate methodology comprising of the benchmarks and performance metrics for each specific phase of the analysis is the key to successfully answering the question like: (a) why is a system under-utilizing the available hardware? (b) why isn't the system faster on the new server? (c) are the new processors more energy efficient?

The first question one needs to answer is which type of benchmarks is the most beneficial for a certain analysis. For example, main-memory optimized OLTP systems have been proliferating recently with the falling costs of main-memory and increasing demand for higher transactional throughput. Such systems forgo many components of the traditional designs, such as the buffer pool, and feature lightweight concurrency control mechanisms, cache-conscious indexes and optimized query compilation techniques. The natural question to ask is whether they manage to utilize the modern processors more efficiently than their disk-based predecessors. Macrobenchmarks from TPC family offer a good starting point as their behavior is well understood in the community and there is a wealth of previous data. By using metrics such as the number of instruction retired per cycle (IPC) and the percentage of stall cycle we can gain broad understanding of the behavior of different systems. In this instance, despite all the design differences, in-memory OLTP behaves very similarly to the traditional systems; stalling over 60% of the time while running the TPC-C workload.

While great for deriving broad insights, macrobenchmarks are not able to help in "what-if analysis" due to their inflexibility. For such scenarios, microbenchmarks are

© Springer International Publishing AG 2017
R. Nambiar and M. Poess (Eds.): TPCTC 2016, LNCS 10080, pp. 159–160, 2017.
DOI: 10.1007/978-3-319-54334-5

much better choice. For example, one can measure the impact of the amount of time spent inside the OLTP engine on the micro-architectural behavior by using a simple micro-benchmark where every transaction reads/writes N number of rows per transaction, N being $1, 10, 100, 500, \ldots$ In this case, while disk-based systems exhibit better use of micro-architectural resources, in-memory systems suffer higher percentages of stall cycle as the amount of time spent inside the OLTP engine increases. Another scenario where microbenchmarks are essential is in quantifying the impact of non-uniform topologies of modern multisocket multicore servers. Nowadays, such servers feature Hardware Islands, i.e., groups of cores that communicate fast among themselves and slower with other groups. By using partition-sensitive microbenchmark containing the single site and multisite transaction and varying their ratio, we can effectively compare different deployment configurations of the distributed OLTP systems. We conclude that no single optimal configuration exists: the ideal configuration depends on the hardware topology and the workload.

While macro/microbenchmarks are useful for understanding the system behavior from different aspects, they are only meaningful based on the defined metrics used to analyze the system behavior. Metrics can be used to quantify the aggregate system behavior, e.g., delivered throughput and IPC, as well as the fine-grained components explaining the aggregate system behavior, e.g., execution time breakdown. For example, assuming that majority of the execution time goes to memory stalls when running OLTP workloads, breakdown of memory stalls into the stalls coming from different levels of the cache hierarchy shows that in-memory systems not using aggressive transaction compilation still suffers the most from the L1 instruction stalls similar to disk-based systems. In-memory systems using aggressive transaction compilation, on the other hand, can significantly reduce the L1 instruction stalls. This, however, amplifies the last-level cache data stalls, resulting high percentage of stall cycle.

On the other hand, energy efficiency has become a serious concern over the last decade. Traditionally low-power low-end ARM cores gradually advance to server-grade processors. Having observed that OLTP workloads severely under-utilize the Intel Xeon-like processors, we compare the power, throughput and latency characteristics of the two. We observe that, while Xeon's delivered throughput is 1–3× higher, its consumed power is 3–15× larger than the ARM processor, rendering the ARM processor up to 9× higher energy-efficient. On the other hand, ARM's quantified latency towards to the tail of latency distribution can be up to 11× higher, showing that ARM is less suitable for tail-latency-critical workloads.

In summary, choosing the right methodology is essential in understanding the behavior of hardware and software systems. Macrobenchmarks enable gaining broad insights and compare observed behavior with previous analyses. Microbenchmarks enable variation of specific parameters to reveal fine-grained trends in specific dimensions to deepen the understanding. In addition to simple performance metrics, breakdowns are key to explaining the observed trends, especially when used in conjunction with sensitivity analysis to gain complete understanding of the analyzed behavior. Lastly, as power has become an increasingly important concern, energy efficient OLTP highlights a new set of challenges and extends the design and analysis space of transactional systems.

Author Index

Printed in the United States
By Bookmasters